Biochemical & Medicinal Chemistry Series

Bacteria and antibacterial agents

JOHN MANN

Professor of Organic Chemistry
University of Reading

and

M. JAMES C. CRABBE

Professor of Protein Biochemistry
University of Reading

Copublished in the United States with
UNIVERSITY SCIENCE BOOKS

Spektrum Academic Publishers
33 Beaumont Street, Oxford OX1 2PF

Distributed by
W. H. Freeman at Macmillan Press Limited
Houndmills, Basingstoke, RG21 6XS

Copublished in the United States with
University Science Books,
55D Gate Five Road, Sausalito, California 94965
Fax 415–332–5393

British Library Cataloguing in Publication Data
A catalogue record for this book is available from the British Library.
ISBN 0–7167–4508–9

Library of Congress Cataloging-in-Publication Data
Mann, J.
Bacteria and antibacterial agents / John Mann, M. James C. Crabbe
p. cm. —(Biochemical and medicinal chemistry series)
Includes bibliographical references and index
ISBN 0–935702–91–1 (USA)
1. Antibiotics 2. Drug resistance in microorganisms.
I. Crabbe, M. James C. II. Title. III. Series
RM267.M27 1995   615'.329—dc20   95—7004   CIP

Copyright © 1996 Spektrum Akademischer Verlag GmbH

Set by KEYWORD Publishing Services, London
Printed by Bell and Bain Limited

Biochemical & Medicinal Chemistry Series

Series Editor
JOHN MANN
Professor of Organic Chemistry, University of Reading

Titles in this series

Neuropharmacology

T.W. STONE

An Introduction to Biotransformations in Organic Chemistry

JAMES R. HANSON

Bacteria and Antibacterial Agents

JOHN MANN AND M. JAMES C. CRABBE

Contents

Preface

It is just over 50 years since the clinical efficacy of penicillin was established and in this period thousands of new antibacterial compounds have been discovered. Yet we are now facing a situation where most strains of bacteria are resistant to one or more of these antibiotics, so the quest for new compounds with novel modes of activity continues.

The authors believe that all chemists and biomedical scientists should possess a basic knowledge of the biology of bacteria and how their growth can be controlled by antibacterial agents. This book aims to provide a concise introduction to these subjects. Most of the strategies for bacterial control and the mechanisms of antibiotic resistance are included, and all classes of antibacterial compounds are described, with special attention given to the β-lactam antibiotics. This is not a research text, though a list of leading research references is provided, but it does contain all the key information needed for those who are commencing research projects involving bacteria and their control. It should be suitable for upper-level chemistry undergraduates (who might become involved in the synthesis of antibiotics), as well as for medical students, who will use the drugs described.

We thank Professor Douglas Young for reading the first draft of the book. His comments and suggestions were invaluable. In addition, we are immensely grateful to David Brown, who drew most of the chemical structures.

1 Microorganisms and disease

Introduction

During his studies on fermentation, between 1857 and 1876, Pasteur, conscious of the practical applications of his scientific work, devoted considerable attention to the spoilage of beer and wine, which he showed to be caused by the growth of undesirable microorganisms. Pasteur used a significant term to describe these microbially induced spoilage processes; he called them 'diseases' of beer and wine. In fact, he was already considering the possibility that microorganisms may act as agents of infectious disease in higher organisms. A certain amount of evidence in support of this hypothesis already existed. It had been shown in 1813 that specific fungi could cause disease of wheat and rye, and in 1845 M. J. Berkeley had proved that the great Potato Blight of Ireland, a natural disaster which deeply influenced Irish history, was caused by a fungus. The first recognition that fungi could be specifically associated with an animal disease came in 1836 with the work of A. Bassi in Italy on a fungal disease of silkworms. A few years later, J. L. Schönlein showed that certain skin diseases of man were caused by fungal infections. Despite these indications, very few medical scientists were willing to entertain the notion that the major infectious diseases of man could be caused by microorganisms, and fewer still believed that organisms as small and apparently simple as the bacteria could act as agents of disease.

The introduction of anaesthesia about 1840 had made possible a very rapid development of surgical methods. Speed was no longer a primary consideration, and the surgeon was able to undertake operations of a length and complexity that would have been unthinkable previously. However, with the elaboration of surgical technique, a problem that had always existed become more and more serious—surgical sepsis: the infections that followed surgical intervention and often resulted in the death of the patient. Pasteur's studies on the problem of spontaneous generation had shown the presence of microorganisms in the air and at the same time indicated various ways in which their access to and development in organic infusions could be prevented. A young British surgeon, Joseph Lister, who was deeply impressed by Pasteur's work, reasoned that surgical sepsis might well result from microbial infection of the human tissues exposed during operation. He decided to develop methods for preventing the access of microorganisms to surgical wounds. By the scrupulous sterilization of surgical instruments, by the use of disinfectant dressings, and by the conduct of surgery under a spray of disinfectant to prevent airborne infection, he succeeded in greatly reducing the incidence of surgical sepsis. Lister's procedures of antiseptic surgery, developed about 1864, were initially greeted with considerable scepticism but, as their striking success in the prevention of surgical sepsis was recognized, they gradually became common practice. This work provided powerful indirect evidence for the germ theory of disease, even though it did not cast any light on the possible microbial causation of specific human diseases.

The discovery that bacteria could act as specific agents of infectious disease in higher animals was made through the study of anthrax, an infection of domestic animals that is transmissible to man. The conclusive demonstration of the bacterial causation, or etiology, of anthrax was provided by a German doctor, Robert Koch. While Pasteur could be called the 'father' of microbiology, Koch was an intellectual giant in the field of bacteriology. He proved that the anthrax bacillus was the sole cause of the disease, and demonstrated that its epidemiology was a result of the natural history of the bacterium. This work was published in 1876 and 1877; later he quarrelled with Pasteur over anthrax. Despite his severe myopia, further researches in microscopy and bacteriology led to his discovery on 24th March, 1882 of the tubercle bacillus that causes TB—a major cause of death at that time.

In 1890 he produced tuberculin as a lymph-inoculation cure for tuberculosis, and although it did not prove effective as a cure, it was useful in diagnosis. He became professor at Berlin and director of the Institute of Hygiene in 1885, and first director of the Berlin Institute for Infectious Diseases in 1891. In 1896 and 1903 he was summoned to South Africa to study rinderpest and other cattle plagues. He won the Nobel Prize for physiology or medicine in 1905.

A different approach was made by Paul Ehrlich, a one-time colleague of Koch. Ehrlich proposed that infectious diseases might be curable by the use of drugs selectively toxic to the microorganism, i.e. by selective antimicrobial chemotherapy. Initially, he was interested in understanding the factors that influenced the uptake and distribution of dyestuffs by the various cell types. The vast array of dyes produced by the German chemical industry at that time allowed him to make a thorough exploration of the effects of size, ionic character, and the functional groups present in the molecules, on these biological characteristics. As a result Ehrlich was able to develop new dyes with specific affinities for different cell types, including the first staining procedure for the tubercle bacillus which made use of its acid-fast characteristics. As a result of these experiments he contracted tuberculosis and had to spend 2 years in Egypt in order to recover.

His first major contribution to chemotherapy arose from his work on trypanosomal infections, such as the sleeping sickness caused by infection with *Trypanosoma gambiense* transmitted by the tse-tse fly. He tested over 100 dyes and established the efficacy of the azo dye Nagana Red (**1**) as an anti-trypanosomal agent. This dye is relatively water insoluble and its efficacy could be much enhanced by the addition of a sulfonic acid group to produce Trypan Red (**2**). However, field tests on humans in Uganda resulted in unacceptable side effects resulting in blindness or even death!

Ehrlich's attention was then drawn to a report that the arsenic compound Atoxyl (**3**), which had been developed as an alternative to potassium arsenite (the main constituent of Fowler's solution) as an anticancer agent. He embarked upon a programme of analogue preparation with the avowed aim of producing new drugs with good 'parasitotropic' activity combined with low 'organotropic' effects (i.e. low toxicity towards the host). Today we would say that he sought drugs with a good therapeutic index. The correct structure of Atoxyl, the 4-aminophenyl derivative of arsenic acid, was first established, and then a whole range of analogues were prepared. Many of these are extremely toxic, and the investigators were initially puzzled by the fact that much higher concentrations of the compounds were needed to kill trypanosomes in culture than could possibly be attained in treated mice. Yet clinical cures were achieved. Ehrlich reasoned that some kind of metabolic

$HNCH_2CO_2^{\ominus} \ Na^{\oplus}$

As ... n

4

OH

NH_2

As ... n

5

activation was occurring in the animals, and speculated that the drugs were being reduced *in vivo*. As a result they prepared a range of arsenoxides and arsenobenzenes such as arsenophenylglycine (**4**) and arsphenamin (**5**)—the 606th compound they prepared, and subsequently know as Salvarsan. This last compound had spectacular activity against the protozoa *Treponema pallidum* which causes syphilis, and Salvarsan rapidly became the drug of choice for treating this condition. A single dose of 900 milligrams was sufficient to effect a cure, and in 1910 around 65,000 vials of the drug were provided free of charge for clinical evaluation purposes. This compound was the first true chemotherapeutic drug and Ehrlich is justifiably known as the 'father of chemotherapy'.

Ehrlich believed that Salvarsan and the other compounds had a dimeric structure, but more recent work has shown that they are typically polymeric in nature. Their mode of action almost certainly involves prior metabolism to an arsenoxide which then targets and interacts with thiol groups on key enzymes.

Classical antibiotics: sulfonamides, penicillins and cephalosporins

It should be recalled that during Victorian times the average life expectancy was a mere 45 years and infant mortality (from birth to age 5 years) was 150 per 1000 live births. Many of these premature deaths can be attributed to the life-threatening conditions produced following bacterial infection. Even as recently as 1919, during the great influenza pandemic, more than 20 million people died due to complications (generally pneumonia) caused by bacterial infection rather than the primary viral infection. The fragility of life prior to the advent of antibiotics is thus self-evident.

Sulfonamides

The first hesitant steps that led to the discovery of the sulfonamides and penicillins respectively, occurred almost simultaneously in Germany and England. In 1927, Gerhard Domagk at the laboratories of I. G. Farbenindustrie, began a systematic investigation directed towards the identification of agents effective against haemolytic streptococci. These bacteria are typically associated with throat infections, but may precipitate potentially fatal conditions such as rheumatic fever, acute nephritis, meningitis and pneumonia. He worked initially with a strain of streptococcus isolated from a patient who had died from septicaemia, and studied the effects on it of compounds with established bactericidal properties, in particular gold compounds, acridines and azo dyes. Only the latter class of compounds showed consistently good antibacterial activity combined with low toxicity towards the test animals (mice). One red dye, Prontosil rubrum (**6**), which fortuitously contained a sulfonamide group, was chosen for a clinical trial, and in the early part of 1933 the drug was used to save the life of a 10-month old baby dying of staphylococcal septicaemia. The child's recovery rom the disease was little short of miraculous, and this success was subsequently repeated in England when a total of 38 dangerously ill women were treated and only three failed to respond. The drug was immediately marketed and became the first of the family of antibacterials known as the sulfonamides. Domagk received the Nobel Prize for his contributions in 1939, 4 years after his daughter was cured of septicaemia through the use of Prontosil rubrum!

The mode of action of the sulfonamides was elucidated following the discovery that Prontosil rubrum (**6**) was metabolized to 4-aminobenzene sulfonamide (**7**), and that this had similar potency to the parent drug. Indeed, the compound was put on sale in the USA in 1937 under the tradename Elixir of Sulphanilamide, but unfortunately the formulation included diethylene glycol as a solvent. This induced severe liver and kidney damage, and 76 patients died before the drug was withdrawn. One positive effect of this disaster was the creation of a strict Food and Drug Act in the USA, and the American public has been spared the effects of a number of other drugs, most notably thalidomide, due to this enactment.

In 1940, Donald Woods of the Department of Biochemistry at Oxford University, showed that 4-aminobenzoic acid (**8**) could antagonize, i.e. overcome, the antibacterial effects of sulfanilamide, and proposed that the sulfon-

pteridine   glutamic acid

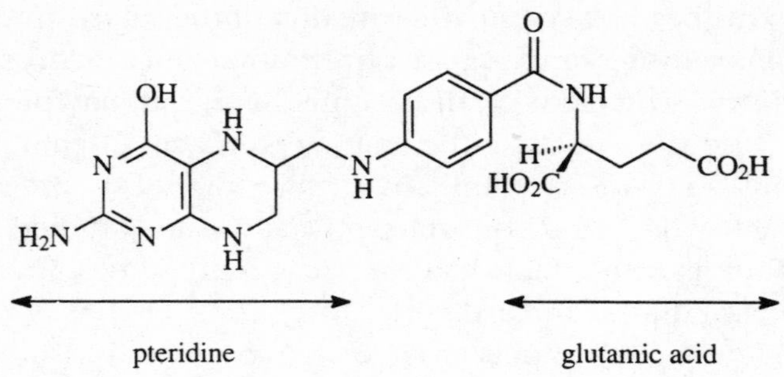

9

amide was acting as an anti-metabolite of 4-aminobenzoic acid. This latter compound was required by bacteria for the biosynthesis of the essential co-factor tetrahydrofolate (**9**) via dihydrofolate, the other constituents being a substituted pteridine and glutamic acid (see Figure 1.1). This co-factor is a key component of most metabolism that involves the transfer of 1-carbon units. The sulfonamide has sufficient structural and electronic similarities to the natural substrate for it to be accepted by the bacterial enzymes as an alternative substrate, but production of tetrahydrofolate is then not possible. Mammals, including humans, can only produce tetrahydrofolate (**9**) if they have a supply of the vitamin folic acid (see Figure 1.1) as a constituent of their diet. The sulfonamides thus have selective toxicity towards many bacteria. The enzyme dihydrofolate reductase is another obvious point for chemotherapeutic intervention, since inhibition of this enzyme would also prevent the production of tetrahydrofolate. This strategy is explored further in Chapter 3.

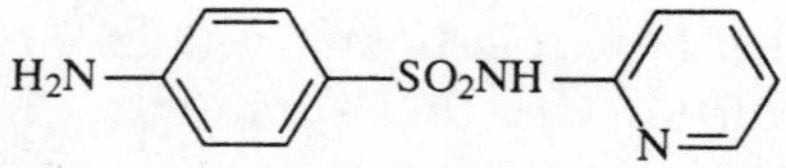

10 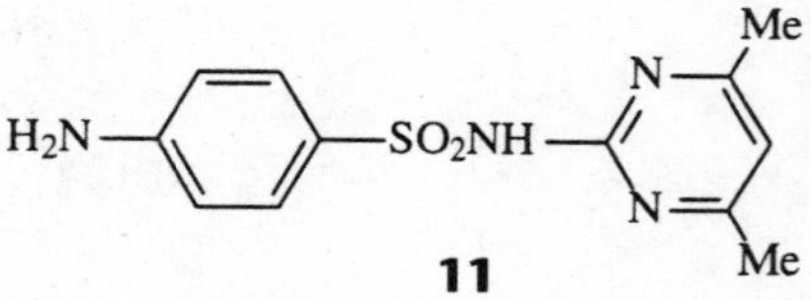   **11**

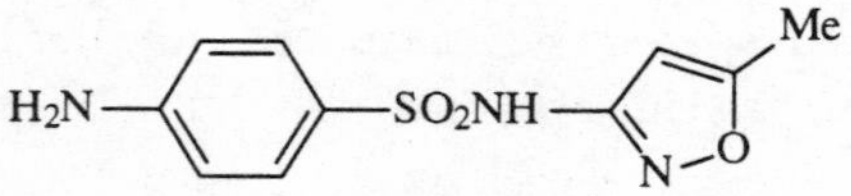

12

Thousands of sulfonamides have been synthesized over the years and sulfapyridine (**10**), sulfadimidine (**11**) and sulfamethoxazole (**12**) have been particularly successful. The first of these attained star status (as May and Baker 693, i.e. M&B 693 or simply M&B to the general public) because it was used to save the life of Winston Churchill who contracted pneumonia on a trip to

North Africa in 1943. The introduction of the sulfonamides in the late 1930s had another major effect—a significant reduction in mortality following childbirth. Most causes of death were the triad of puerperal sepsis (about 40%

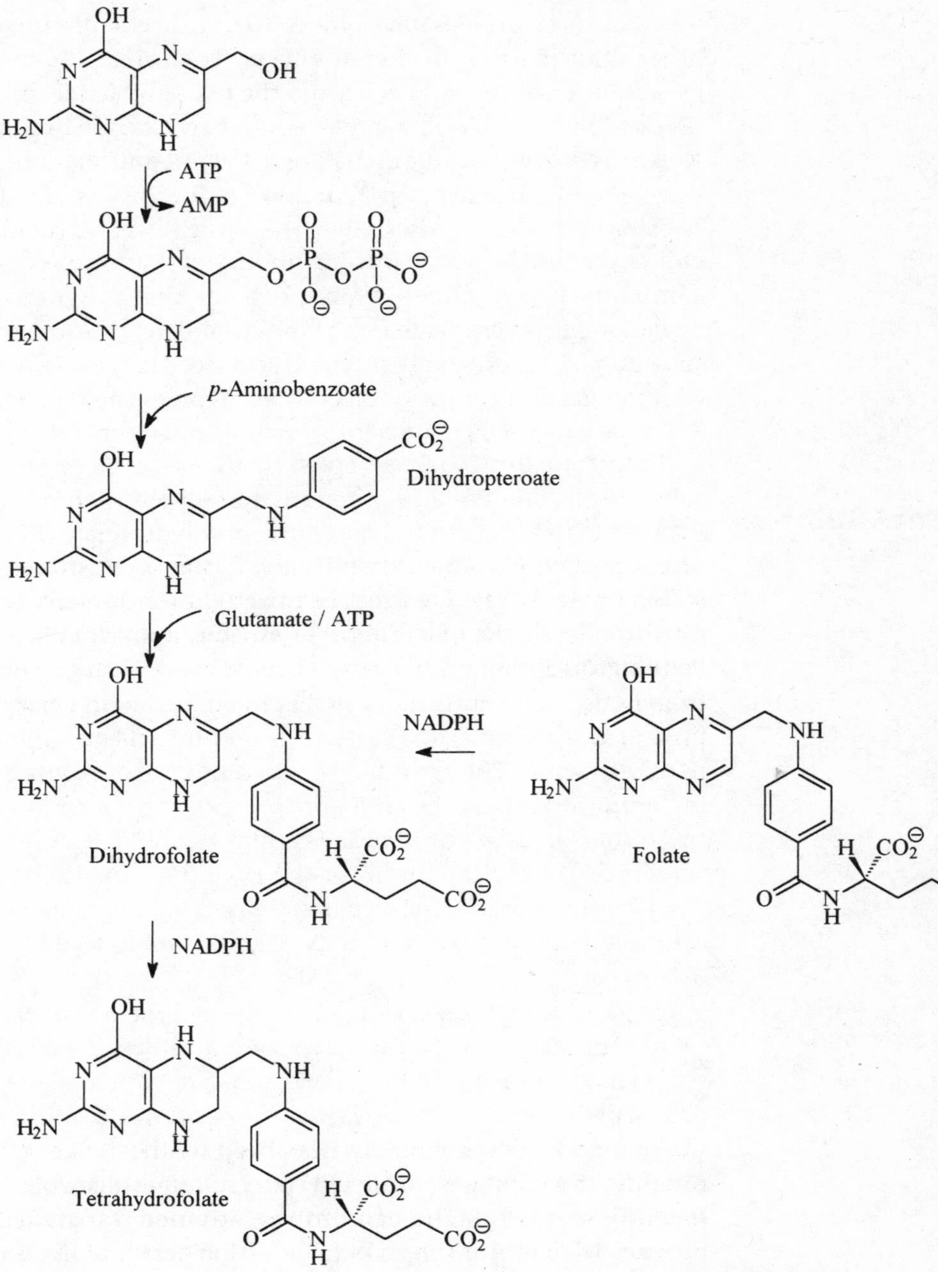

Figure 1.1

of deaths), toxaemia (20–25%) and haemorrhage (20%). The sulfonamides had a major impact on the first of these causes, and the death rate fell from 50 in 10,000 (at the turn of the century) to 5–6 per 10,000 in 1950, and to 1 in 10,000 today.

Other sulfonamides have proved to have useful pharmacological properties other than their antibacterial effects. The uricosuric drug probenecid (**13**) used for the treatment of gout, and the hypoglycaemic drug tolbutamide (**14**), used to lower the blood sugar levels in diabetics, are two important examples. Notwithstanding the dramatic impact of the sulfonamides in the mid-1930s, they are not particularly potent and do not possess broad spectrum activity. These attributes are possessed by the penicillins and cephalosporins.

13

14

Penicillins and cephalosporins

There is an enormous amount of myth associated with the discovery of penicillins by Alexander Fleming, but the generally accepted sequence of events can be told simply. Fleming was a bacteriologist at St. Mary's Hospital in Paddington and spent the early 1920s trying to identify new antiseptic agents from bodily secretions. They had early success with lysozyme (a protein comprising 125 amino acids) isolated from human tears and subsequently from hens' eggwhite. These studies required the use of cultures of various bacteria including streptococci, staphylococci and pneumococci, and it was one of these that became contaminated by the mould *Penicillium notatum* during September 1928. The origin of the mould has never been determined, but whatever its source Fleming correctly noted that in the vicinity of the mould no growth of staphylococcal cells occurred and indeed the bacteria were being destroyed.

Fleming and his coworkers did not have the chemical or biochemical expertise to capitalize on their discovery, and although he published his work and spoke about it on a number of occasions, it was largely overlooked. This was primarily because crude extracts of penicillium moulds, such as *Penicillium glaucum* and *P. brevicompactum* had been used by Lister and others in the late nineteenth century as antibacterial agents, so Fleming's discovery did not seem particularly significant. What was not recognized was the superior potency of Fleming's 'mould juice'—that was left to Charles Florey and his group to demonstrate.

They too began work (around 1935) with lysozyme and in 1938 they investigated the properties of Fleming's strain of *P. notatum*, which had thankfully been sub-cultured continuously by the London group. Florey's group contained a number of excellent chemists and biochemists, most notably Chain, Heatley and Abraham, and they were quickly able to obtain relatively large amounts of highly active and relatively pure (by the purification standards of the time) penicillins. Appropriate and highly successful animal tests were carried out, and in the Spring of 1941 they began their first clinical trials. Six patients were treated, all of whom were suffering from life-threatening conditions that had failed to respond to chemotherapy with sulfonamides. Of the six, two died and four survived. Despite these tragic deaths, the survival of the others was little short of miraculous.

These momentous events were taking place at a time when an imminent German invasion was feared, and there was no enthusiasm or finance obtainable from the British pharmaceutical industry. So, with the permission of the Medical Research Council, Florey and Heatley left for the USA with a sample of their culture to investigate the possibility of large-scale production in the United States. A consortium of American drug companies began experiments, initially using the Fleming strain of *P. notatum* and surface culture techniques, but later using deep fermentation processes (similar to those used in beer making) and a more potent strain, *P. chrysogenum* that had been discovered growing on a rotten canteloup in Peoria market, Illinois. By the end of 1944 the American companies were producing well over 100,000 million units of penicillin per month, and there were ample supplies to satisfy the ever-increasing demands of the military.

The actual chemical structure of the major penicillin, benzyl penicillin (**15**) or penicillin G, was in doubt until 1943 and was not finally confirmed until Dorothy Hodgkin concluded her X-ray structural work in 1945. These studies revealed the unexpected presence of the beta-lactam ring fused to a thiazolidine ring, and a number of laboratories accepted the challenge of trying to create this unusual bicyclic system. John Sheehan of Merck was the first to succeed, in 1957, with a synthesis of phenoxymethylpenicillin (**16**) (penicillin V) and this work also allowed obtention of 6-aminopenicillanic acid (6-APA) (**17**). This had long been a compound of interest, because it allowed the construction of a whole range of semi-synthetic penicillins through the addition of side-chains of almost any type.

This synthetic triumph was somewhat overshadowed by the discovery at Beechams in 1958 that 6-APA was in fact always present as a stable product of fermentation, and could be isolated in quantity from the fermentation medium. This revolutionized the synthesis of semi-synthetic penicillins, and over the last four decades literally thousands of novel penicillins have been produced. Of these, ampicillin (**18**) and amoxycillin (**19**), with their good mix of stability and broad spectrum activity, have been particularly widely used. The pro-drug form of ampicillin, pivampicillin (**20**), is also of special interest since this overcomes the somewhat poor absorption of ampicillin from the gut. The ester is more easily absorbed from the gut and the relatively acid-labile ester is then hydrolysed in the bloodstream to yield ampicillin.

The ready availability of penicillins in the 1950s led to a degree of over-prescribing, but more importantly to their widespread use in agriculture. The bacteria thus came into contact with these novel compounds at a much earlier stage and in much greater amounts than they might have otherwise done. Not surprisingly many types of bacteria were able to develop resistance, and by the mid-1960s only the newer semi-synthetic penicillins were of any use against serious bacterial infections.

This situation was partially alleviated by the advent of the cephalosporins, which had a much broader spectrum of activity and a lower tendency to induce resistance in bacteria. Their discovery followed the isolation of a mould, *Cephalosporium acremonium*, from a sewage outlet in Cagliari, Sardinia in 1948. Extracts of this mould were shown to inhibit the growth of typhus bacilli and other pathogens in culture, but were also given to patients with typhoid and paratyphoid, with positive results. Florey and his group were sent a culture of the mould and during the 1950s they isolated and characterized a number of antibacterial components, but only cephalosporin C (**21**) had the

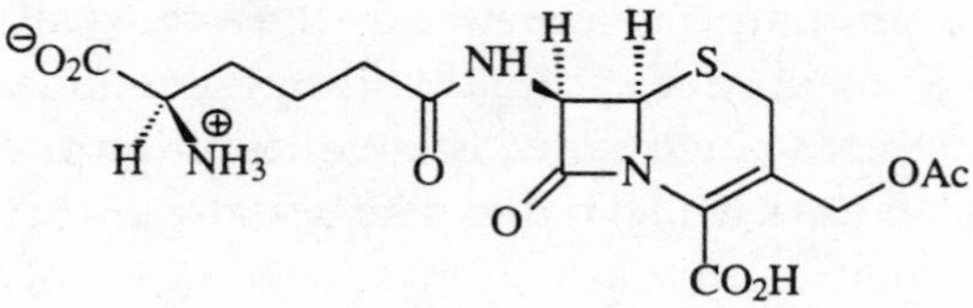

21

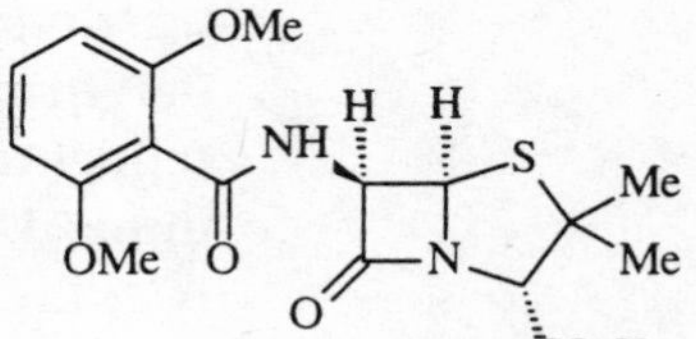

22

desired mix of broad spectrum activity and resistance to the array of destruct-ive enzymes, e.g. penicillinase, produced by the resistant bacterial strains. The introduction of methicillin (**22**) which had good stability towards peni-cillinase, by Beechams in 1960 appeared to negate any advantage that cephalosporin C possessed, and the planned commercial launch of the anti-biotic was abandoned.

This situation was neatly reversed when the Eli Lilly company discovered a remarkable piece of chemistry. Abraham and Newton at Oxford had shown that cephalopsorin C possessed a similar structure to the penicillins, but contained a dihydrothiazine ring rather than a thiazolidine ring. Morin and co-workers at Lilly showed that the ring system of penicillin G could be converted into the ring system of cephalosporin C (see Figure 1.2). This one-stage chemical

Figure 1.2

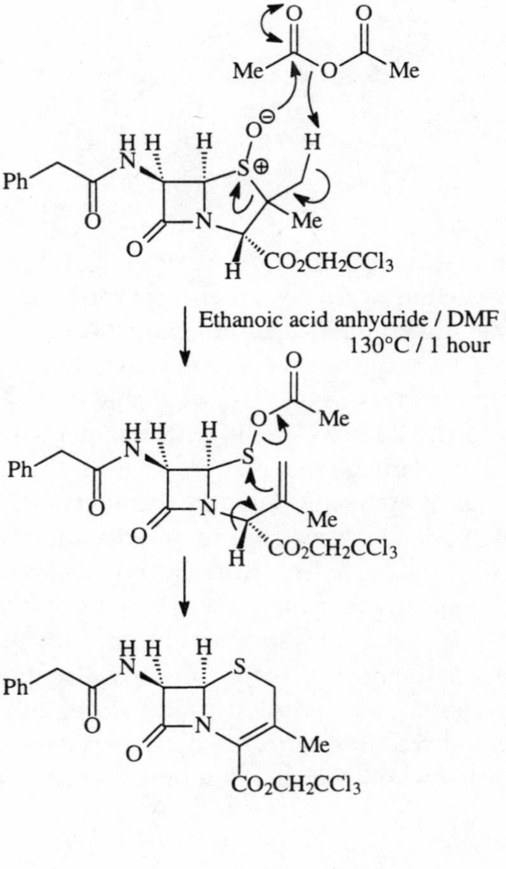

Ethanoic acid anhydride / DMF
130°C / 1 hour

transformation, coupled with new methods for the removal of the phenylacetyl side-chain (Figure 1.3), opened the way for the production of a plethora of new cephalosporins. Of the hundreds of antibiotics since produced, one should mention cephalexin (**23**), which has to date been the most widely used.

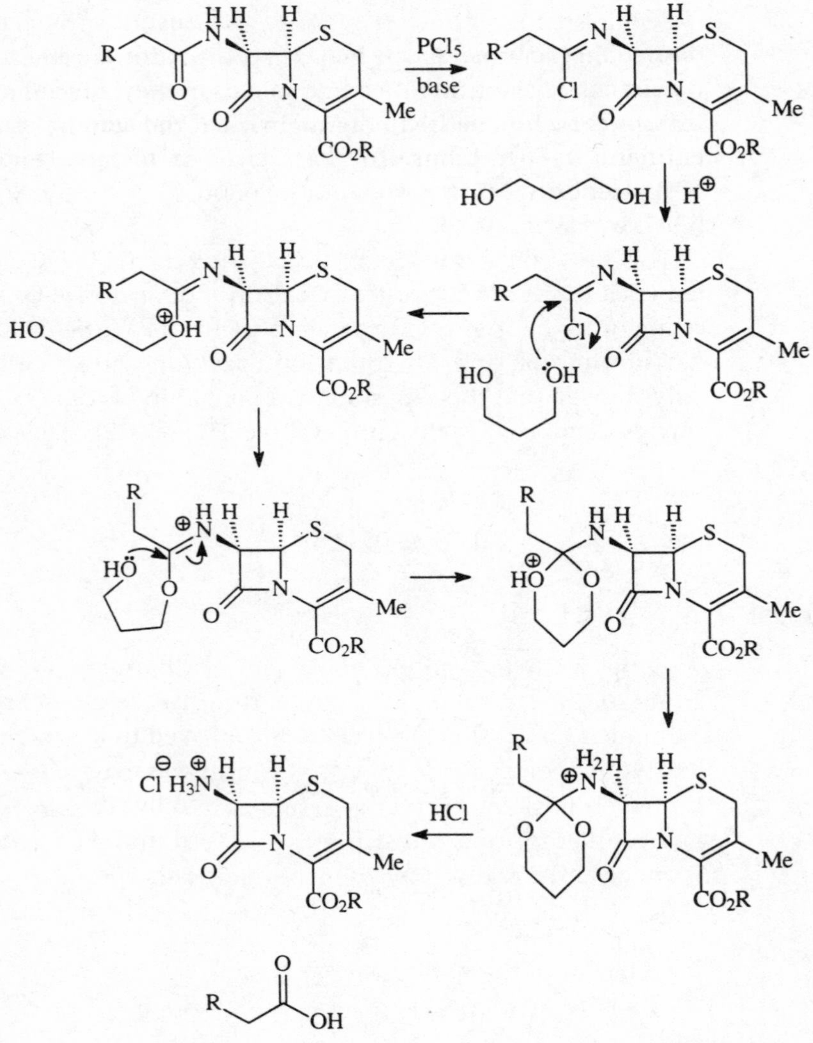

Figure 1.3

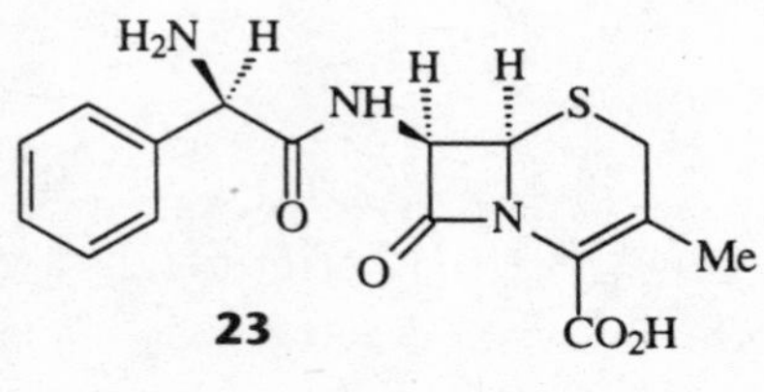

All the other major classes of antibiotics were discovered as part of screening programmes involving soil microorganisms. The American microbiologist Selman Waksman believed that the disappearance of pathogenic bacteria from the soil (e.g. soil contaminated with sewage) was caused primarily by the activities of the various soil microorganisms. In 1943 he commenced a major screening programme to identify organisms with activity against *Mycobacterium tuberculosis*, the causative agent for tuberculosis. After screening thousands of strains of actinomycetes, they identified one, *Streptomyces griseus*, which possessed potent antibacterial activity. The major active constituent was streptomycin (**24**), the first member of a family of aminoglycoside antibiotics.

24

Other discoveries followed soon after. Chloramphenicol (**25**) was isolated from *Streptomyces venezuelae*, an actinomycete discovered in a soil sample from near Caracas in Venezuela. This proved to be the most broad spectrum antibiotic known at the time (1947), and was even active in the treatment of typhus. It has now been largely superseded by the semi-synthetic penicillins and cephalosporins, but still proves useful in the treatment of typhoid, salmonella and certain forms of meningitis.

25

The tetracyclines (Figure 1.4) were also discovered as part of a screening programme with actinomycetes, and chlortetracycline (*Streptomyces aureofaciens*), oxytetracycline (*S. rimosus*), and tetracycline (*S. viridifaciens*) are the

Figure 1.4

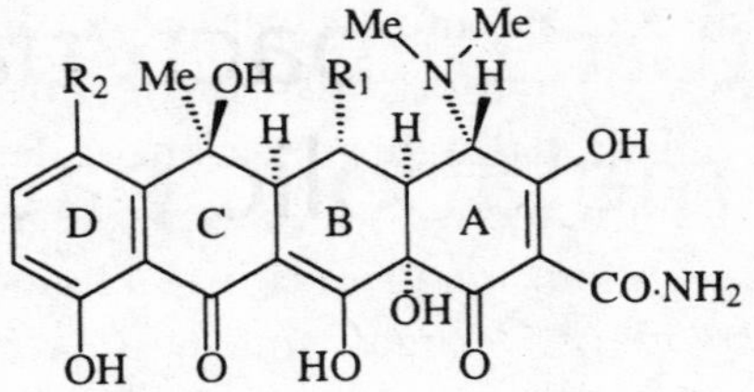

Tetracycline $R_1 = R_2 = H$
Chlorotetracycline $R_1 = H$ $R_2 = Cl$
Oxytetracycline $R_1 = OH$ $R_2 = H$

most important antibiotics in this class. These compounds had broad spectrum activity and could be taken orally, though their poor absorption in the gut led to almost inevitable disturbance of the gut bacterial flora with resultant side effects. Bacterial resistance was also apparent at an early stage.

Finally, another chlorine-containing antibiotic, griseofulvin (**26**), was isolated following problems associated with a tree-planting programme on Wareham Heath in Dorset. The trees failed to thrive and the soil was subsequently shown to be almost devoid of typical soil microorganisms. The culprit was shown to be the mould *Penicillium janczewskii* and the antibacterial agent, griseofulvin, was shown to be identical to a compound previously isolated from *Penicillium griseofulvum*. The compound proved to be particularly effective as an antifungal agent, and is now usually reserved for the treatment of ringworm in both humans and animals.

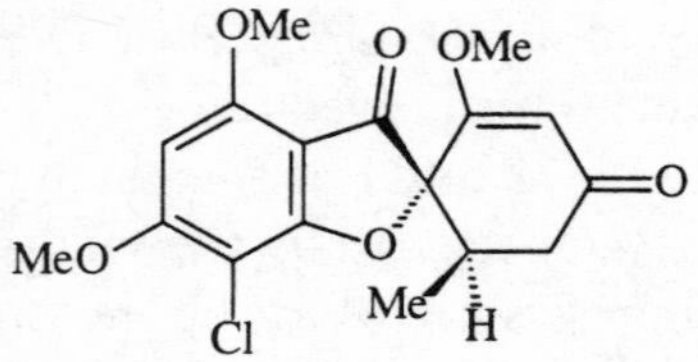

26

The modes of action of these antibiotics and a discussion of the newer antibiotics will be the subject of later chapters. The next two chapters will deal with the basic biology and biochemistry of bacteria.

2 | The structure of bacteria and their metabolic pathways

Introduction

This chapter provides a summary of the biochemical events and structures that underlie bacterial growth. For a model system, we can consider an unspecified bacterium that can grow in a medium containing glucose as the carbon source. Its nitrogen requirements are satisfied by ammonium ions and its sulfur requirements by sulfate. Magnesium and phosphate are essential and it needs trace amounts of other metals (e.g. iron). It can be considered as a 'generalized bacterial cell' and we shall attribute to it a mixture of the properties found in several different kinds of bacteria. It should be regarded as an abstraction in much the same way as the average man.

The organism is represented schematically in Figure 2.1. It is rod-shaped and has a rigid outer wall that maintains and supports the membrane that it encloses. The wall is made of a polymeric substance, the mucopeptide, and the membrane contains proteins and lipids. These coats surround the cytoplasm, which consists mainly of polymers: deoxyribonucleic acid (DNA);

Figure 2.1

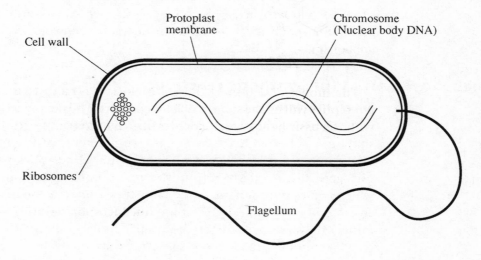

ribonucleic acid (RNA); proteins and polysaccharides. In terms of dry weight the polymers account for about 90% of the cell. The remaining 10% of the cell is made up of a large variety of small molecules: amino acids, nucleotides, vitamins and fats. Although these constitute so small a fraction of the cell mass, they are metabolically of great importance.

Not only do the macromolecules make up the bulk of the bacterial cell, they also give it the characteristics which distinguish it from all other types of bacteria. The small molecules, on the other hand, are largely common to all types of bacteria and, indeed, to other forms of life.

When the organism is in a suitable environment or growth medium, more of all these materials is produced and in due course the cell divides into two daughter cells indistinguishable from one another and from their parent. The number of chemical reactions involved in growth is unknown, but is probably of the order of many thousand. Of these, a few may occur spontaneously, but the vast majority have to be catalysed by specific proteins, the enzymes. Each of these catalyses a specific reaction such as the addition or removal of water, or hydrogen, or 1-C residues, or amino groups, etc. We may consider enzyme reactions being classed as either catabolic or anabolic.

Catabolic reactions (degradative reactions)

There is, first, a complex series of enzymes which degrade glucose to smaller aliphatic carbon compounds. These are the degradative enzymes. The net process is exergonic, i.e. produces energy. It also results in the supply of carbon skeletons for synthetic reactions. The three major degradative pathways are: glycolysis for the degradation of glucose to yield pyruvate and thence acetate (as acetyl thioCoenzyme A); hydrolysis of proteins to their constituent amino acids; and hydrolysis of lipids to produce glycerol (1,2,3-trihydroxy-propane) and fatty acids which may be further oxidatively degraded to yield acetyl thioCoenzyme A. These are the major building blocks of biosynthesis or anabolism.

Anabolic reactions (biosynthetic reactions)
Small molecules

From these skeletons a further series of enzymes catalyses the formation of the small molecules which are the basic components of the macromolecules. Many of these intermediates (amino acids, nucleotides, hexosamines) contain nitrogen which is derived from the ammonium salts. Some contain sulfur which comes from sulfate. At the same time some small molecules (vitamins, co-factors) are synthesized but are not incorporated into macromolecules; rather, they are needed for the proper functioning of the enzymes. The enzymes producing all these substances may be called biosynthetic. As a group they largely require energy and are therefore endergonic. The energy is produced by the catabolic reactions.

Macromolecules

A further series of enzymes then converts the basic small molecules into macromolecules. When enough of these have been synthesized the cell divides. Since the distinctive character of the cell is determined by its macro-molecules, much of this book is concerned with the mechanisms of inhibition of the series of reactions that produce these compounds.

The genetic information for copying the cell is carried in its DNA which is the 'blueprint' for the whole cell, that is, all the information determining what the biochemical machinery shall be and how it will be put together is encoded in the DNA. When the cell divides each of the daughter cells must, apart from anything else, receive a complete copy of the 'blueprint'. It is essential that the DNA molecules should be copied correctly at every division because, as in our highly organized system, a random error will almost certainly be damaging. It is only very rarely that such an error will be advant-ageous. However, advantageous mutations do occur, and have led, among other things, to bacterial resistance to β-lactam antibiotics.

Growth and the regulation of biosynthesis

We can now summarize the events that take place when some viable cells are placed into growth medium. Some of their enzymes degrade glucose, some synthesize basic molecules and yet others assemble macromolecules, includ-ing more of all the enzymes. With more of all these catalysts thus available, the same processes will continue but at an accelerated rate, giving yet more enzymes, and more cells. The rate of synthesis is thus proportional to the amount of cell material present and this leads to an exponential rate of growth that continues until something in the environment becomes limiting. When this happens some types of bacteria simply cease to grow but others form spores which are heat-resistant and can lie in a dormant state for many years. Subsequently, if the environment becomes favourable, these can ger-minate and begin to grow again. Sporulation and germination are among the most primitive forms of cell differentiation and are consequently of consider-able interest.

Returning to the actively growing cell, let us consider its internal economy. It has to produce 20 types of amino acids for its proteins, four types of nucleotides containing deoxyribose for DNA, four more containing ribose for RNA and also a variety of co-factors and lipids. The synthesis of any one of these substances may easily involve ten or more specifically catalysed steps carried out by biosynthetic enzymes. In addition there is a considerable number of intermediates produced from glucose by the catabolic enzymes.

For efficient growth all the basic materials and all the macromolecules derived from them have to be produced in the correct proportions. Under natural conditions, bacteria are probably often in competition for a limited amount of nutrient. The consequence is that a very efficient regulation mechanism has evolved. Since virtually all metabolic steps are enzymatically catalysed, in considering metabolic regulation, we have really to consider regulation of enzymatic function. There are two ways in which this can occur. One is by altering the amount of any particular enzyme, the other is by altering the rate at which it functions. Both types of regulation are found in the bacterial cell and their combined effect ensures that the cell is geared to get maximum yield of protoplasm from its environment and to do so in the minimum time.

The bacterial cell: major structures

Bacterial cells occur in all sorts of different shapes and sizes depending on the kind of organism and on the way in which it has been grown, but for many purposes it is possible to disregard these variations and to consider the common properties of the 'generalized bacterial cell' (Figure 2.2). Thus, although some bacteria are spherical or curved or spiral, the majority are rod-shaped and are about 1000 nm wide and 2000 nm in length. A single bacterial cell may thus have a volume of 10^{-12} ml and contain 2.5×10^{-13} g of dry matter

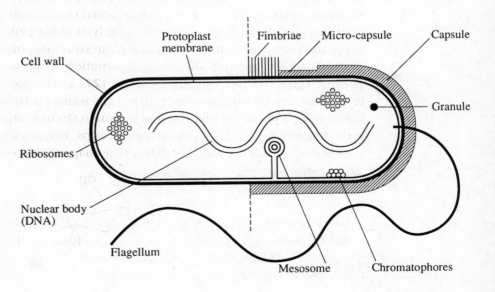

Figure 2.2

(equivalent to a molecular weight of 1.5×10^{11}). The bacterium, being a prokaryote, does not contain a nucleus to house its DNA. Eukaryotes, yeasts fungi and more 'advanced' organisms, including mammals, have nuclei to contain DNA and mitochondria, where reactions involved in respiration take place. But this bacterium is not just an undifferentiated blob of 'protoplasm'. It is a highly organized structure with some organelles corresponding in function to many of those found in higher organisms. The hereditary material (DNA) is embedded in the cytoplasm which, surrounded by the cell membrane, is called the protoplast. Outside this lies the cell wall.

Cell walls and membranes

With a few exceptions, the cells of prokaryotic organisms are enclosed within walls. In itself, this is not a distinctive group character, as cell walls also exist in many eukaryotic protists and in plants. However, analytical studies have shown that the prokaryotic cell wall has a unique chemical composition. In all biological groups, the cell wall is chemically complex, being typically composed of a number of different kinds of macromolecules, associated either in a molecular mosaic or in adjacent layers. However, the tensile strength which enables the wall to withstand the osmotic pressure of the enclosed protoplast and hence prevents osmotic lysis in a hypotonic environment is largely contributed by one particular component of the wall, which forms a molecular mesh, or sacculus, enclosing and constraining the protoplast and conferring on it the characteristic external form of the cell. The nature of this component can often be inferred by the effect of exposing cells to hydrolytic enzymes, known to destroy one specific macromolecular constituent of the wall. If such treatment results in swelling and osmotic lysis of the cell, it is clear that the constituent in question is essential for the structural integrity of the wall. Thus, the cells of higher plants undergo osmotic lysis upon treatment with the enzyme cellulase, pinpointing cellulose (**27**) as the constituent primarily responsible for the structural integrity of the wall. Similarly, the polysaccharide, chitin, (**28**) can be identified as the main strengthening element of the wall in many fungi. In all prokaryotic groups, the sacculus, which confers tensile strength on the wall and defines the shape of the cell, is composed of a

27

28

unique kind of organic polymer, known as murein. Mureins are heteropolymers, composed of several kinds of subunits, some of which are carbohydrates and some amino acids. Bacteria are classically divided into Gram-positive and Gram-negative organisms, according to their reaction to the Gram stain. This is an empirical procedure where the cells are treated successively with the dye crystal violet and iodine, decolourised and treated with safranine. Gram-positive cell walls contain little lipid, but high concentrations of teichoic acids, and are less complex than Gram-negative cell walls.

Investigations have proceeded along two main lines: firstly, the electron microscope used in conjunction with thin sectioning has revealed details of surface structures and appendages in a way hitherto impossible (Figure 2.3); and secondly, biochemical investigations of the whole cell or isolated parts of it have shown the chemical material of which the individual structures are composed. An attempt is frequently made to relate the microscopy information obtained from the study of sections with the biochemical results obtained from studies of cell fractions and so to build up a coordinated picture of the structure in the intact cell.

Not all the structures revealed by the electron microscope and by other means appear to be necessary for survival of the bacterium. By comparing different kinds of bacteria it is possible to arrive at the lowest common denominator in terms of those that are necessary for growth and division. These 'essential' structures are depicted diagrammatically in Figure 2.2, together with other structures which are found in some, but not all, bacteria. In electron micrographs (Figure 2.3) the cytoplasm has a granular appearance resulting from the presence of a large number of ribosomes. Within the cytoplasm

Figure 2.3

the nuclear area chromatinic body, bacterial nucleoid can be seen in thin sections. The nucleoid has a very fine fibrillar network but unlike the animal or the plant cell nucleus is not separated from the cytoplasmic contents by a nuclear membrane.

In a large number of bacteria, membranous structures termed mesosomes or chondrioids are found closely associated with the nucleoid and with the cytoplasmic membrane at the site of septation. Various granules (lipid, glycogen, volutin or metaphosphate) may be found in the cytoplasm, and in photosynthetic bacteria large numbers of membranous structures termed chromatophores can be seen: these are the organelles in which photosynthesis occurs.

The cell wall may be covered with fimbriae or pili but little is known of the function of these thin threads. Motile bacteria have at least one flagellum, the flagella being very long thin structures composed of protein subunits. Many bacteria have a capsule external to the cell wall and this structure is sometimes many times thicker than the bacterium itself.

The wall is thus fairly rigid and gives shape and protection to the cell. It amounts to about 10% of the weight of the entire cell. A mucopeptide is always present, and this seems to be what makes the wall rigid. The mucopeptides are made of chains of amino sugar, N-acetyl-glucosamine (**29**) alternating with N-acetylmuramic acid (**30**). To form the basic unit of the oligo-saccharide (Figure 2.4) short peptides are linked to the muramic acid residues and separate chains may be joined by these peptides to form the two-dimensional structure needed for a wall (Figure 2.5). More importantly the bacterial enzyme transpeptidases catalyse the cleavage of the terminal D-alanine residue and cross-linking to the

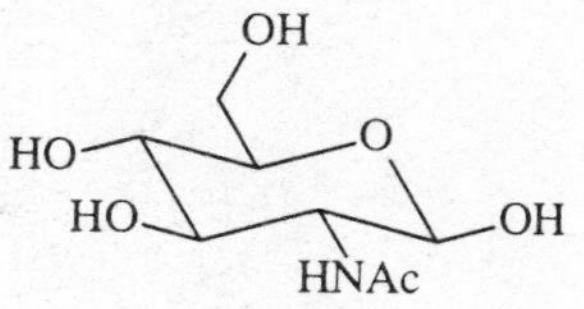

Figure 2.4

29

30

N-Acetyl muramic acid
NAMA

N-Acetylglucosamine
NAG

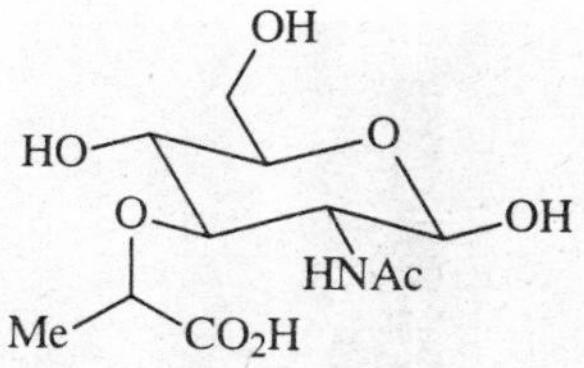

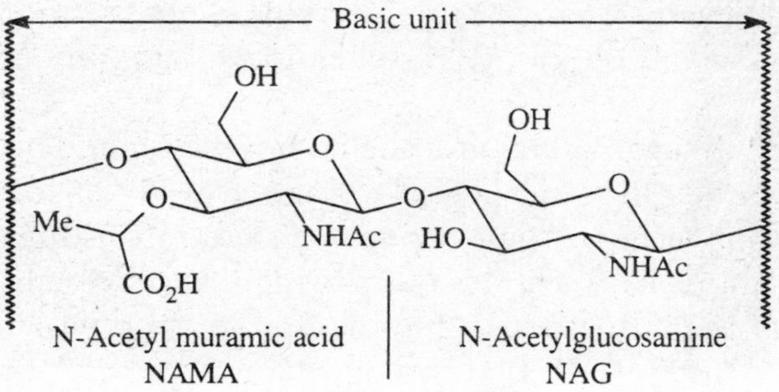

```
—GINAc—MurNAc——————GINAc—          —GINAc—MurNAc——————GINAc—
        |                                   |
       L-ala                               L-ala
        |                                   |
       D-glu                               D-glu
        |                                   |
       L-lys—gly-gly-gly-gly-gly-NH₂        L-lys—gly-gly-gly-gly-gly-NH₂
        |                                   |
      (D-ala)₂                             D-ala
                                            |
                                           D-ala
```

Figure 2.5

```
—GINAc—Mur—NAc—GINAc—               —GINAc—MurNAc——————GINAc—
        |                                   |
       L-ala                               L-ala
        |                                   |
       D-glu                               D-glu
        |                                   |
       L-lys—gly-gly-gly-gly-gly-NH         L-lys—gly-gly-gly-gly-gly-NH₂
        |                                   |
      (D-ala)₂                             D-ala
                                            +
                                           D-ala
```

Figure 2.6

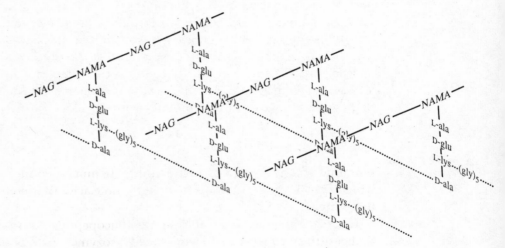

side-chain glycine of another polypeptide chain to produce a 3-dimensional array as shown in Figure 2.6. The chemistry of this process is shown in Figure 2.7.

The amino sugar, muramic acid, has not been found in any biological polymers other than the mucopeptides of the cell walls of bacteria and the closely related cyanobacteria. These peptides are also interesting in that they contain unusual amino acids. Besides L-alanine (**31**) they contain D-alanine (**32**) and D-glutamic acid, the so-called 'unnatural' isomers which are not present in proteins. Most species also contain diaminopimelic acid (**33**) which, like

Figure 2.7

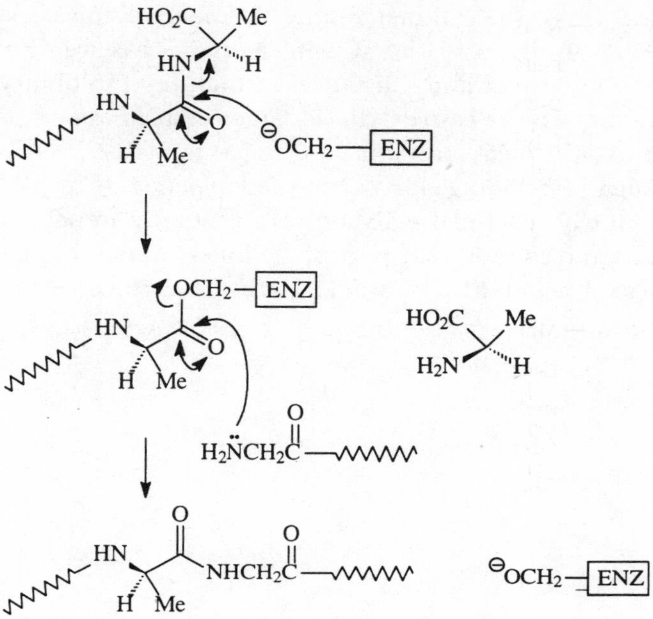

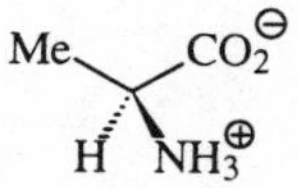

                                    31 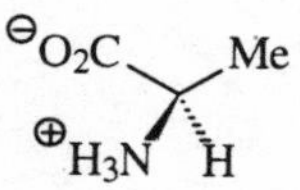   **32** 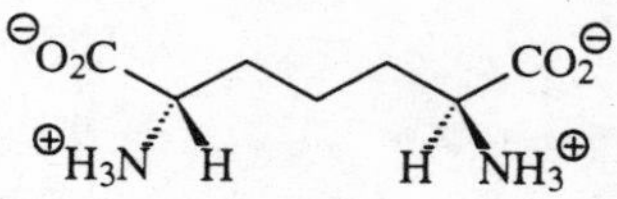   **33**

muramic acid, is restricted in nature to these muco-peptides. Other polymers which may occur in cell walls include teichoic acids, lipopolysaccharides and lipoproteins.

In species in which the wall is simply mucopeptide it is sometimes possible to digest the wall with an enzyme called lysozyme which occurs in secretions such as tears and sweat and also in white of egg. Enzymic digestion of the wall releases the protoplast but this is likely to burst unless given some osmotic protection. This is because the concentration of intra-cellular solutes produces an effect equivalent to 5–20 atmospheres pressure. The protoplast is bounded by a membrane called the plasma membrane, the protoplast membrane, or the cytoplasmic membrane. This structure consists predominantly of protein and lipid, with a thickness of about 8 nm. The membrane is the main permeability barrier of the cell since the wall is freely penetrated by most molecules except very large ones. Some substances pass into and out of cells by passive diffusion but many are transported by highly specific systems which require energy and are located in the cell membrane. The name permease, or sometimes translocase, is given to such a system. As far as passive diffusion is concerned, smaller molecules and substances of high lipid solubility penetrate through membranes more easily than do larger molecules and polar substances. For instance, 4-C sugars and some 5-C sugars may pass freely but other 5-C and all 6-C sugars (including glucose) may fail to penetrate by passive diffusion except very slowly. Bacterial cells are also generally impermeable to some small cations and to inorganic phosphate ions. These non-penetrating substances have to be actively transported into the organisms.

3 | The sites of action of antibacterial drugs

Introduction

Antibacterial drugs exert their action by interfering with either the structure or the metabolic pathways of bacteria. Important methods of antibiotic action include interfering with metabolic pathways, binding to the cytoplasmic membrane, inhibiting protein biosynthesis, inhibiting nucleic acid biosynthesis, and disrupting cell wall biosynthesis. This chapter will deal briefly with these five modes of action.

Interference in metabolic pathways

The importance of the enzyme co-factor tetrahydrofolate was mentioned in Chapter 1, its deprivation has wide-ranging effects on bacterial cells that inhibit the biosynthesis of either amino acids, purines or pyrimidines. The production of this co-factor is inhibited by sulfonamides, which act as competitive inhibitors of the enzyme dihydropteroate synthetase. Bacterial metabolism can also be selectively disrupted by trimethoprim (**34**) which inhibits bacterial, but not eukaryotic, dihyrofolate reductase. Bacterial resistance to sulfonamides has been countered by using a mixture of the sulfonamides, sulfamethoxazole (**35**) and trimethoprim. Sulfonamides inhibit the incorporation of p-aminobenzoic acid (PABA) into folic acid, while trimethoprim prevents the reduction of dihydrofolate into tetrahydrofolate; the latter form is essential in 1-carbon transfer reactions. Selectively is ensured as mammalian cells utilize preformed folates from the diet, and do not synthesize the compound, while trimethoprim is active against bacterial, but not mammalian, dihydrofolate reductase. The drug combination sulfamethoxazole and trimethoprim is sold under a variety of brand names, including Septrin. *p*-Aminosalicylic acid, PAS, (**36**) acts by a mechanism similar to that of sulfonamides and is active against mycobacteria. It can be used as an antituberculous agent.

The arsenical compound Salvarsan (**37**), long used for the treatment of syphilis and sleeping sickness, owes its toxic side effects to the reaction of its oxidation product arsen oxide with free thiol groups on proteins.

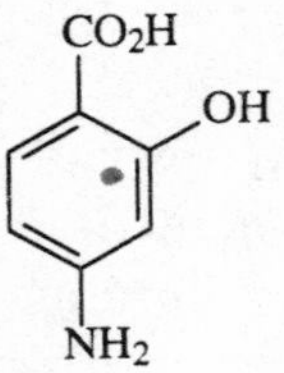

36 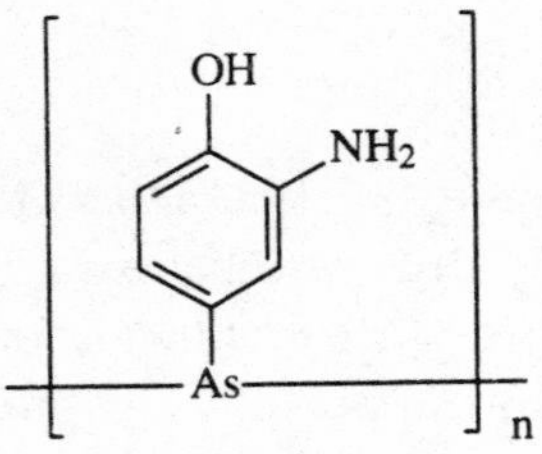   37

Binding to the cytoplasmic membrane

Many compounds have bactericidal action by binding to cytoplasmic membranes. Some cause major disorganization and loss of function of the membrane. Others, by insertion into the membrane, form a pore which alters its permeability to ions. Examples of the first group of compounds include detergents and certain phenolic compounds such as hexachlorophene, (**38**). These compounds form the basis of commercial antiseptic preparations which have useful bactericidal properties when applied to the external surfaces of the body. Some cyclic polypeptide antibiotics, such as tyrocidin A, gramicidin S (**39**) and polymixins also bind to the cytoplasmic membrane whereas others such as valinomycin (**40**), monensin A, amphotericin B and nigericin (stack) in the membrane forming pores. Valinomycin, gramicidin and amphotericin are neutral ionophores, i.e. compounds that can sequester an ion, while nigericin

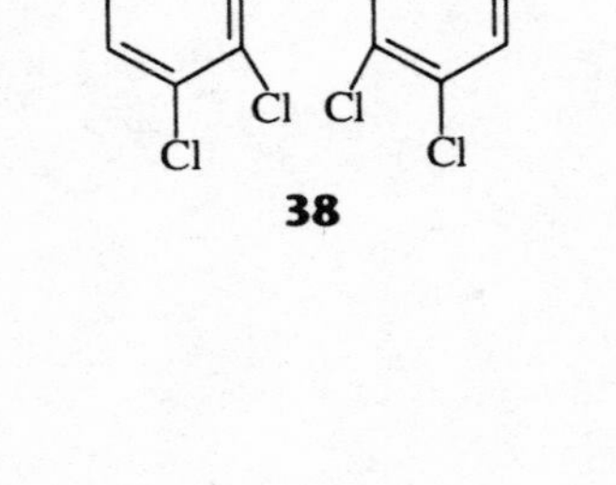

38

Gramicidin S
S₁ X = Y = L-Val
S₂ X = 2-Aminobutanoic acid, Y = L-Val
S₃ X = Y = 2-Aminobutanoic acid

39

40   D-Hyi = D-Hydroxyisovaleric acid

(41) and monensin A (42) have an acidic ionizable group. Nigericin and monensin A are so-called macrolide antibiotics and act as ionophores with the selective permeabilites:

$K^+ > Na^+ > Li^+$ (for nigericin)

$Na^+ > K^+ > Rb^+; Li^+$ (for monensin A)

41

42

Tyrocidin A (43) and gramicidin S are cyclic decapeptides which cause damage to the bacterial cell membrane with resultant loss of amino acids and phosphate from the cell. The polymixins are a family of cyclic heptapeptides which induce disorganisation of the cell membrane with resultant disruption of permeability control.

43

Amphotericin B (44) and nystatin (45) are structurally related polyene antibiotics. After formation of the pores they induce leakage of ions and small molecules from sensitive bacterial cells. These include K^+, Na^+, Rb^+, Li^+, phosphate, carboxyl and amino acids, erythritol, arabitol, ribose, sorbose, xylose, urea and phosphate esters.

Valinomycin (40), a cyclic dodecapeptide, causes a dramatic increase in the permeability of biological membranes to potassium, but not sodium; the structure has an almost perfect fit for a potassium ion. Its lipophilic side chains ensure solubility in the cell membrane.

44

45

The principal interactions between amphotericin B and the membrane involve hydrophobic bonds between the lipophilic heptane segment of the antibiotic and the sterols. Such pores have a radius of less than 0.4–0.5 nm. Since these polypeptide antibiotics interact in a similar fashion with the membranes of the host organism they are of limited therapeutic value although monensin is widely used for the treatment of coccidiosis, a protozoal infection in chickens.

Inhibition of protein biosynthesis

Bacterial ribosomes differ in protein composition from those of the eukaryotic host, and there has also been divergence in the sequence of the ribosomal RNAs. Selective inhibitors of ribosomal function constitute clinically useful antibiotics, and inhibitors which abolish similar functions in both bacteria and eukaryotes have also proved useful in dissecting the component reactions of protein synthesis. The process of protein synthesis may be conveniently viewed in three stages: initiation, elongation, and release (Figure 3.1). Firstly DNA is transcribed to provide messenger RNA (m-RNA) which carries the blueprint for protein biosynthesis in the form of codons (trinucleotides). These are

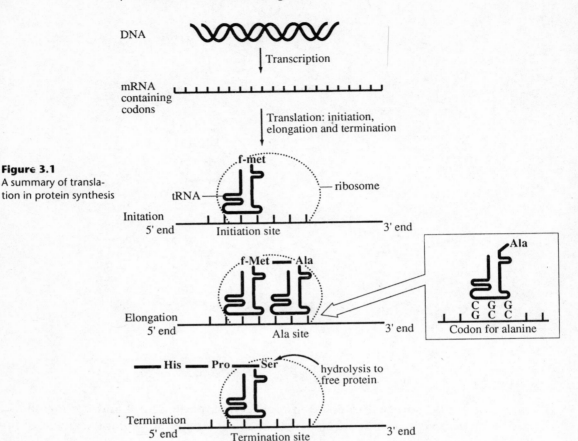

Figure 3.1
A summary of translation in protein synthesis

specific for particular amino acids and the codons subsequently associate with trinucleotide anticodons that are carried by the transfer-RNAs (t-RNAs) which have that particular amino attached at one end. They deliver the correct amino acids when the correct codon-anticodon match is made. During initiation, the m-RNA binds to the ribosome and the unique initiator t-RNA then associates with this part of the ribosome. Elongation involves the sequential addition of amino acid units to the growing polypeptide chain with the order dictated by the codons on m-RNA. Finally, the complete polypeptide (protein) is released by hydrolysis.

The antibiotic streptomycin (**46**), one of the aminoglycosides, has a target site within the 30S subunit and is inhibitory both in initiation and elongation, causing mistranslation of mRNA. It binds to the 30S ribosomal subunit and prevents it from undergoing the normally reversible transition from active to inactive state. After intact bacteria are exposed to streptomycin, polysomes become rapidly depleted, and 70S particles, the 'streptomycin monosomes' build up. Although the formation of the initiation complex is not affected, the complex formed in the presence of the antibiotic cannot synthesize protein and remains fixed in position. Gentamycin, neomycin and kanamycin also may bind to bacterial ribosomes at different sites, causing misreading of messenger RNA (mRNA), and thus mistranslation. They may also be inhibitory to translocation. Kasugamycin (**47**) inhibits a step in initiation.

Tetracyclines (see Figure 1.4) inhibit aminoacyl transfer RNA binding. They bind to both the 30S and 70S ribosomal subunits, and inhibit the binding of aminoacyl transfer RNAs (tRNAs) at the so-called A site. The thiostreptons (**48**) inhibits both aminoacyl tRNA binding and translocation,

R=—NHC(=CH₂)CO·NHC(=CH₂)CO·NH₂

R=—NH₂

by binding to a specific sequence in the 23S RNA of the 50S ribosomal subunit. It also binds to the 70S subunit, but not to the 30S subunit, and inhibits a number of processes in which cytoplasmic protein factors bind to ribosomes. These factors are involved in the initiation, elongation and termination of protein biosynthesis.

Clarification of the mechanism of peptide bond formation on ribosomes was intimately associated with elucidation of the mode of action of puromycin (**49**). It is a structural analogue of aminoacyl adenosine—the 3′ terminus of aminoacyl transfer RNA. Puromycin inhibits elongation in both bacterial and host eukaryotic ribosomes by mimicry of aminoacyl tRNA. The peptidyl–puromycin so formed rapidly dissociates from the ribosome, inhibiting protein synthesis. Puromycin will accept a nascent peptide chain, but since it binds only weakly to ribosomes, the resultant peptidyl–puromycin molecule is soon dislodged from the ribosome.

49

Inhibition of nucleic acid biosynthesis

Bacterial RNA polymerases are composed of two alpha subunits, one beta and one beta′ subunit as the core enzyme. An additional component, the sigma subunit, is required to associate with the core enzyme for selective and high-affinity binding at bacterial promoters. The rifampicins e.g. rifampicin A (**50**) and streptovaricins e.g. streptovaricin A (**51**), members of the ansamycins, specifically inhibit DNA-dependent bacterial RNA polymerase, but not those of the eukaryotic host, binding to the beta subunit. Their site of action is at the level of initiation of RNA synthesis. Streptolydigin, while also binding to the beta subunit, acts by inhibition of chain elongation. Drugs such as 3′-deoxyadenosine (**52**) act as chain terminators of RNA synthesis. Nalidixic acid (**53**) acts upon DNA gyrase, an enzyme specific to bacterial cells, which is

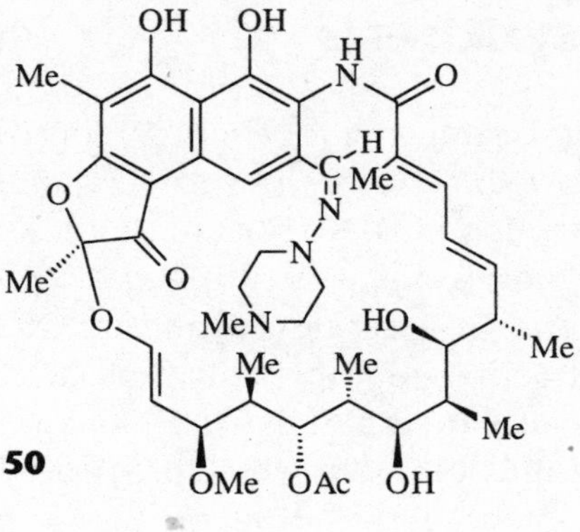

50

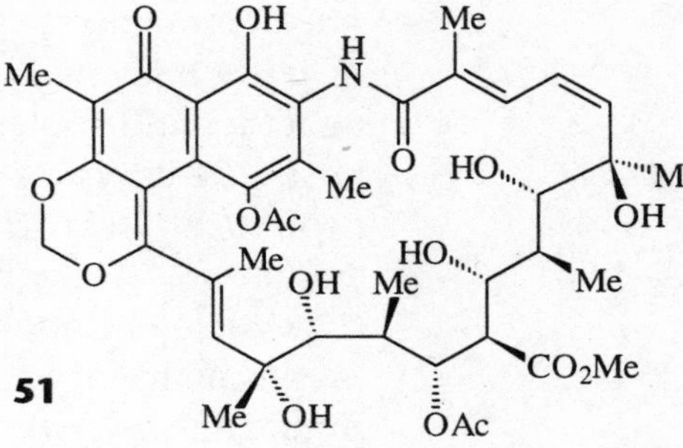

51

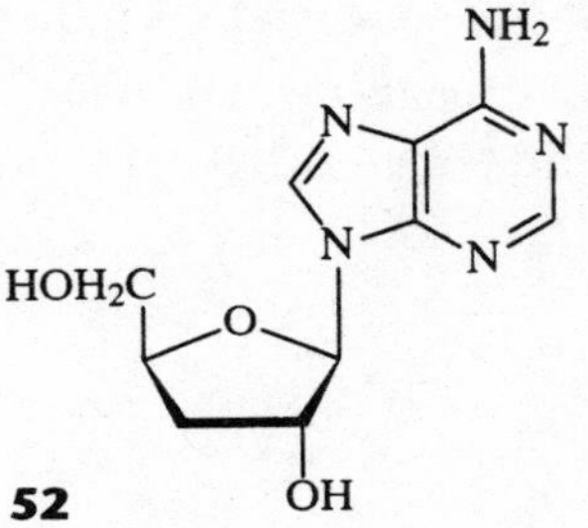

52

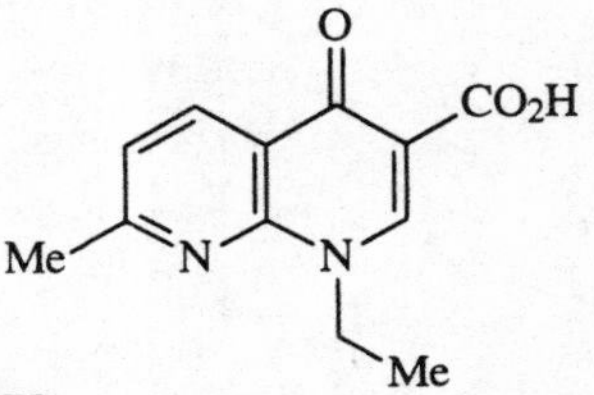

53

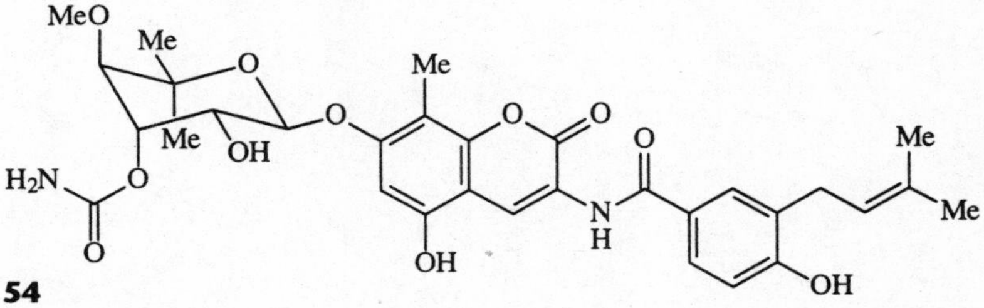

54

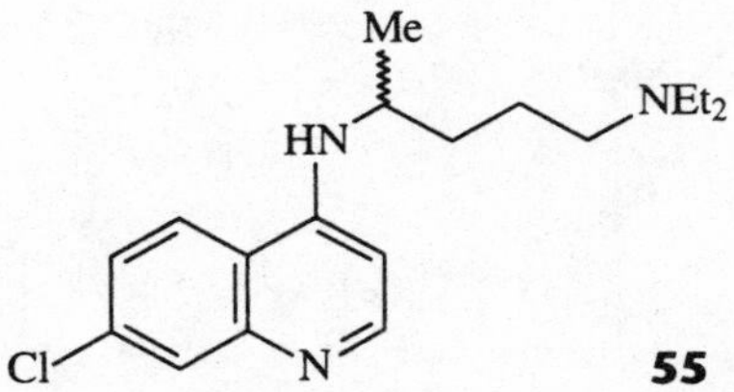

55

involved in unwinding superhelical DNA. Novobiocin (**54**) has similar effects. Hydroxyphenylazopyrimidines are specific inhibitors of the DNA polymerase III of Gram-positive bacteria.

Chloroquine (**55**) is a planar compound that intercalates between the base pairs of DNA, thus distorting the double helix. This compound is still one of the main drugs used in anti-malarial chemotherapy.

Disruption of cell wall biosynthesis

The peptidoglycan layer, accounting for about 5% of the cell wall in Gram-negative bacteria, and 30–40% of Gram-positive cell walls represent a major target for anti-bacterial compounds. No corresponding structure exists in the host. The β-lactam antibiotics (i.e. penicillins and cephalosporins) inhibit peptidoglycan biosynthesis. Cycloserine (**56**) and vancomycin (**57**) act in the cytoplasm, while bacitracin and the penicillins act in the membrane. Cycloserine inhibits the production of D-alanyl-D-alanine by inhibiting the biosynthetic enzyme alanine racemase, which produces D-alanine from L-alanine, and the enzyme D-alanine: D-alanine synthetase, which produces the dipeptide D-alanyl-D-alanine, which in normal growth is used in the synthesis of a UDP–acetylmuramyl pentapeptide (see Figures 2.4 to 2.6). It is assumed that the conformations of D-cycloserine and D-alanine are similar when binding to the enzymes. They both bind covalently to pyridoxal phosphate, the prosthetic group of alanine racemase.

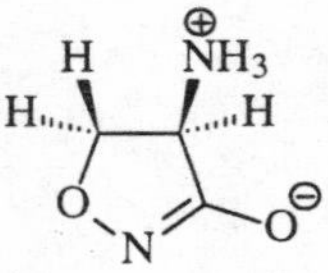

56

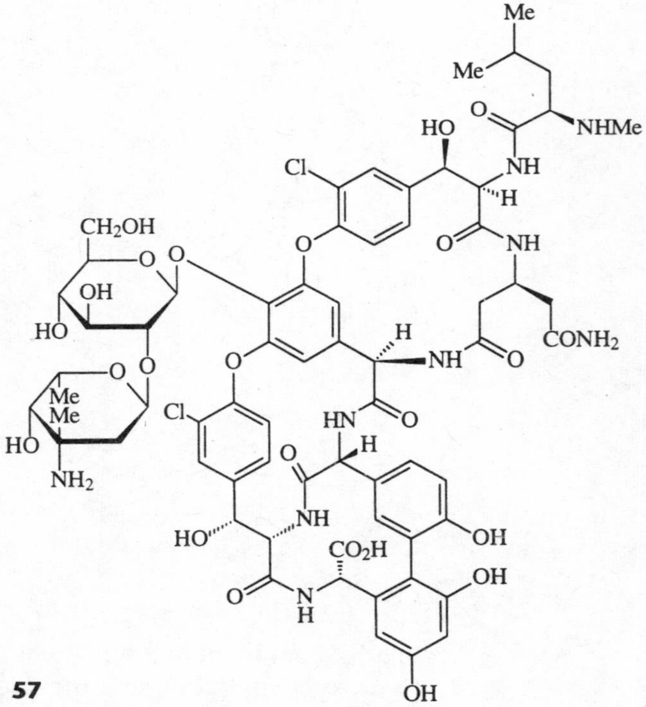

57

Vancomycin and ristocetin (**58**) are members of a family of complex polypeptide antibiotics. They form a complex with the substrate D-Ala-Ala, as shown in Figure 3.2. Nuclear magnetic resonance studies have revealed the existance of three H-bonds between vancomycin and substrate, and five similarly located H-bonds between ristocetin and substrate.

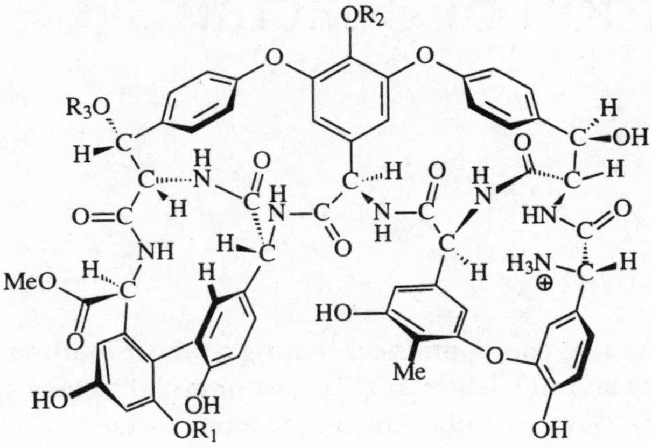

R$_1$ = mannose
R$_2$ = tetrasaccharide
R$_3$ = ristosamine

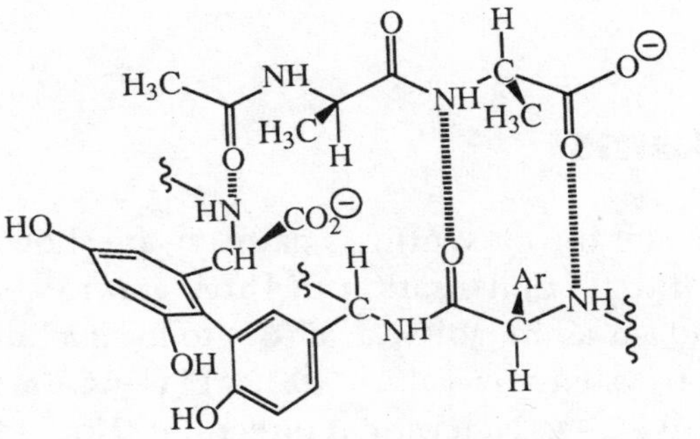

58

Figure 3.2

   Inhibition of peptidoglycan biosynthesis by β-lactams has been the most successful antibacterial strategy since the discovery of the compounds. They have also been the most investigated of the antibiotics, their role in the inhibition of cell wall biosynthesis is discussed in the next chapter.

4

Biosynthesis and mode of action of β-lactams

Introduction

β-Lactams are compounds containing a planar four-membered ring, usually joined to another, larger ring. They comprise the chemical classes of penams, cephems, clavams, carbapenems and monobactams, as shown in that order in Figure 4.1.

Figure 4.1

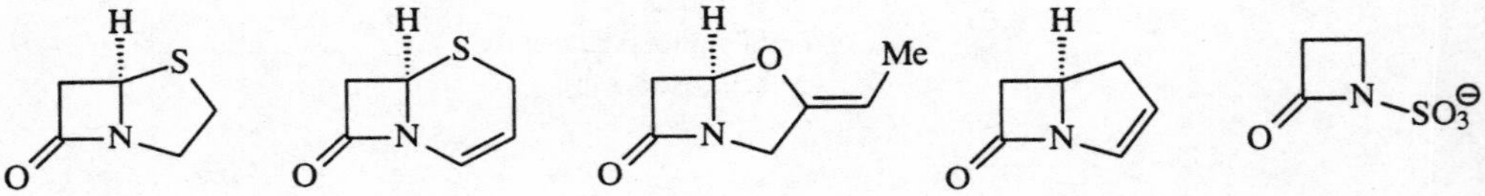

Biosynthesis of β-lactams

The major details of the biosynthesis pathway are shown in Figure 4.2a and 4.2b. After an initial condensation of three amino acids, L-cysteine (**59**), L-valine (**60**) and L-α-aminoadipic acid (**61**) to form a linear tripeptide δ-(L-α-aminoadipoyl)-L-cysteinyl-D-valine (**62**) (LLD–ACV) which is probably synthesised in the C– to N–terminal direction. (Note the change in stereochemistry of the valine moiety.) A single enzyme, isopenicillin N synthetase (IPNS) catalyses oxidation (desaturation) of the linear tripeptide to form the bicyclic isopenicillin N (**63**). The reaction requires ferrous iron, molecular oxygen, and ascorbate. The failure of early attempts to demonstrate the biosynthesis of the penicillin or cephalosporin ring system in cell-free preparations can be attributed, at least in part, to a lack of knowledge of the co-factors required and to the instability of the enzymes concerned in crude extracts of mycelium. However, in a concentrated extract obtained by the lysis of protoplasts of *Cephalosporium acremonium* LLD–ACV was found to be converted with its skeleton intact into a compound which behaved as either penicillin N or isopenicillin N. Later a similar experiment revealed that the

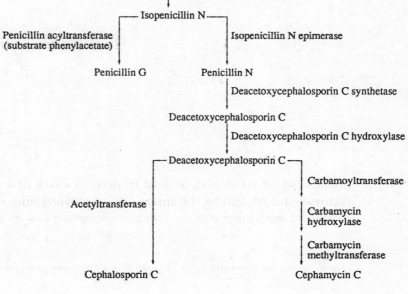

Figure 4.2a

Figure 4.2b

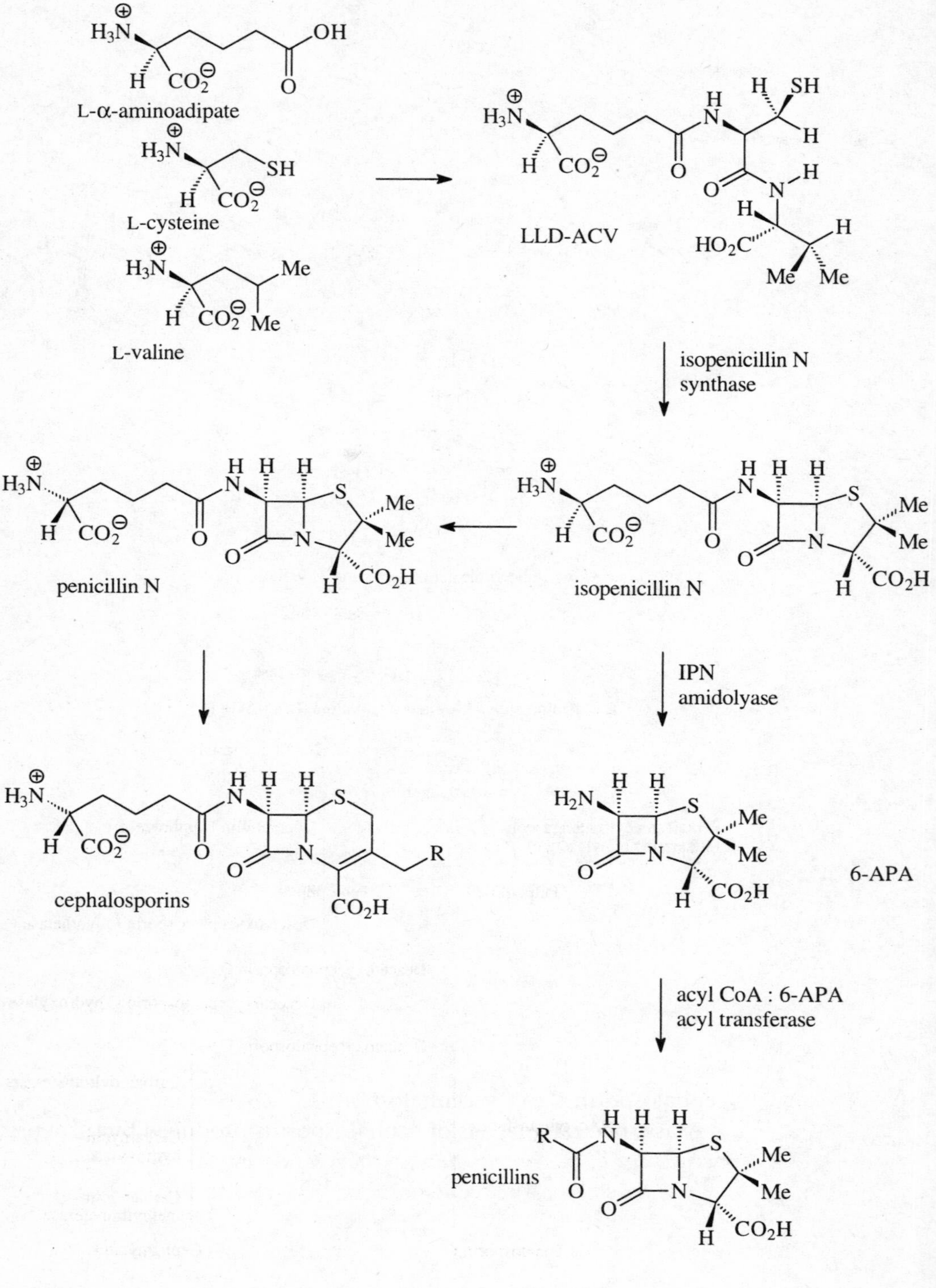
L-α-aminoadipate
L-cysteine
L-valine
LLD-ACV
isopenicillin N
synthase
isopenicillin N
penicillin N
IPN
amidolyase
cephalosporins
6-APA
acyl CoA : 6-APA
acyl transferase
penicillins

product was in fact isopenicillin N and that little or no epimerisation of the α-aminoadipyl group had occurred under the conditions used. The conversion of LLD–ACV to isopenicillin N was also shown to occur in extracts of *Streptomyces clavuligerus* obtained by ultrasonic treatment. Altogether, four different oxygenases each requiring ferrous ions are involved in the biosynthesis of penicillins and cephalosporins. Although the mechanisms of these reactions are not known, work with isotopically labelled substrates has revealed the details shown in Figure 4.2b. A detailed discussion of the presently proposed chemistry catalysed by isopenicillin N synthase is beyond the scope of this book, but the interested reader is referred to the publications (see Further Reading section) from the Baldwin group for information about these fascinating transformations. The group has recently reported the crystal structure of the enzyme which will be important for revealing further information about the mechanisms of these enzyme-catalysed reactions.

For the production of cephalosporins, the five-membered thiazolidine ring is expanded to a six-membered dihydrothiazine ring through an oxidative reaction. This is followed by a hydroxylation reaction. In the fungus *Cephalosporium acremonium* both reactions are catalysed by a single monomeric enzyme. The first reaction in the branch of the pathway that leads from isopenicillin N to cephalosporins in *Cephalosorium* and *Streptomyces* species is catalysed by an epimerase. The penicillin N is partly excreted into the culture fluid and partly converted to deacetoxycephalosporin C. The epimerase was shown to be present in a lysate of protoplasts of *C. acremonium* (and later in one obtained by mechanical disruption of the mycelium) by measurement of the large increase in activity against *Salmonella typhi* which accompanies the isopenicillin N to penicillin N conversion. The epimerase in these extracts was extremely unstable.

Transacylases are involved in reactions at the end of the biosynthetic pathway to cephalosporins, as they are in the case of the hydrophobic penicillins produced by *Penicillium chrysogenum*. An acetyl group is transferred from acetyl CoA to the hydroxyl group of deacetylcephalosporin C in extracts of *C. acremonium* and a carbamoyl group is transferred from carbamoyl phosphate to the hydroxyl group of deacetylcephalosporin C in homogenates of *Streptomyces clavuligerus*. The specificity of the enzyme, or enzymes, involved in these transacylations does not appear to have been established. Other enzymes, the oxgenases, function on the pathways from deacetoxycephalosporin C to the cephalosporins.

Mass spectral analysis of cephalosporins produced by suspensions of the mycelium of *C. acremonium* and *Streptomyces clavuligerus* in the presence of $^{18}O_2$ showed that the oxygen atoms attached to the exocyclic methylene of cephalosporin C and to the β-lactam ring of a 7α-methoxycephalosporin, respectively, were derived from molecular oxygen. It was then found that the oxidation of the methyl group at C-3 in deacetoxycephalosporin C was brought about by a dioxygenase (hydroxylase) requiring Fe^{2+}, ascorbate and α-ketoglutarate for activity and subsequently that the incorporation of a 7α-methoxy group into cephalosporin C (Figure 4.2b) or into the carbamoyl derivative of deacetylcephalosporin C depended on a similar dioxygenase and the transfer of a methyl group from *S*-adenosylmethionine.

Acylase and Transacylase enzymes in *Penicillium* species

Synthetic β-lactams are often made from a precursor molecule, 6-aminopeni-cillanic acid, (**64**). This requires a specific acylase. Reports on the presence of an acylase in *P. chrysogenum* which removes the side-chain of benzypenicillin (**65**) to yield 6-aminopenicillanic acid (6-APA) (See Figure 1.3 for the production of this compound by chemical cleavage of the side chain of penicillin G) and of an acyltransferase which catalyses an exchange of side chains between benzylpenicillin and other solvent-soluble penicillins, or between such penicillins and 6-APA, have a long history. The production of penicillins from 6-APA and acylcoenzyme A in the presence of an acyltransferase was described in 1968 and later a phenylacyl:CoA ligase was isolated. It was suggested that the acylase and transferase activities were associated with a single enzyme. The acylase was said to bring about no detectable hydrolysis of penicillin N or isopenicillin N, but it was shown later that an enzyme, or enzymes, in an extract obtained by grinding the mycelium with sand would catalyse the formation of benzylpenicillin from isopenicillin N, as well as from 6-APA, in the presence of phenylacetyl CoA. The enzyme appeared to have a molecular weight of less than 40,000. It catalysed the formation of 6-APA from both isopenicillin N and benzylpenicillin at an initial rate which was about 30% of the corresponding rate with phenoxymethylpenicillin as a substrate. Since the L-α-aminoadipyl side chain of isopenicillin N was liberated as α-aminoadipic acid the latter was presumably used again in the synthesis of LLD–ACV. However, under the conditions used the reaction virtually ceased after only a small proportion of substrate had been hydrolysed. This could not be attributed simply to a reversible reaction, because no formation of a penicillin was detected when the enzyme was incubated with 6-APA and L-α-aminoadipic acid. But 6-APA itself was found to be a powerful reversible inhibitor of the acylase.

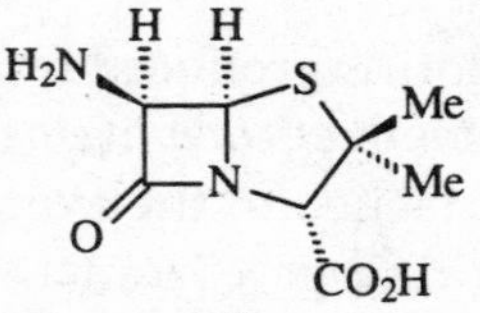

64 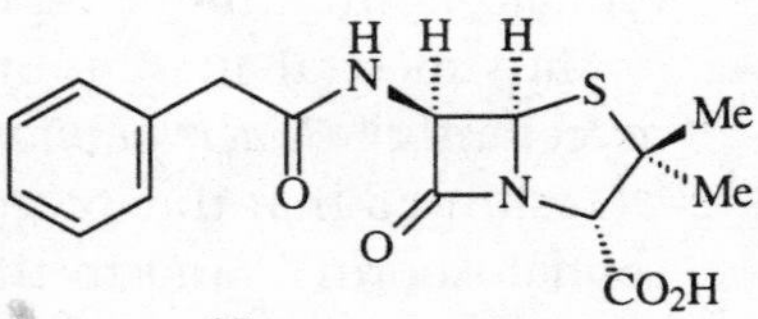   **65**

It has been reported that extracts of several mutant strains of *P. chrysogenum* contain an acyltransferase which can bring about a detectable exchange of the δ-(D-α-aminoadipyl) side chain of cephalosporin C (**66**) for a phenoxyacetyl side chain and thus yield a cephalosporin analogue of penicillin V (**67**).

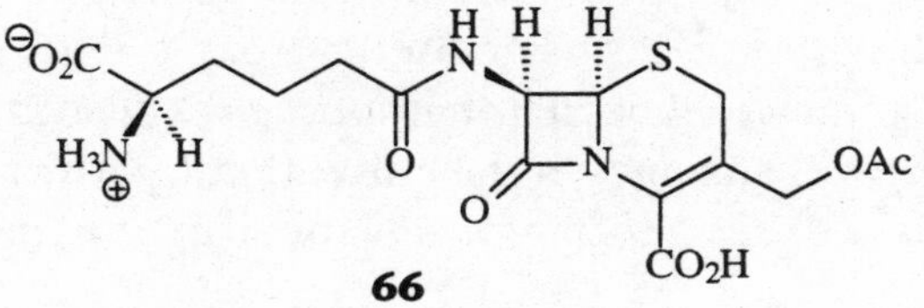

66

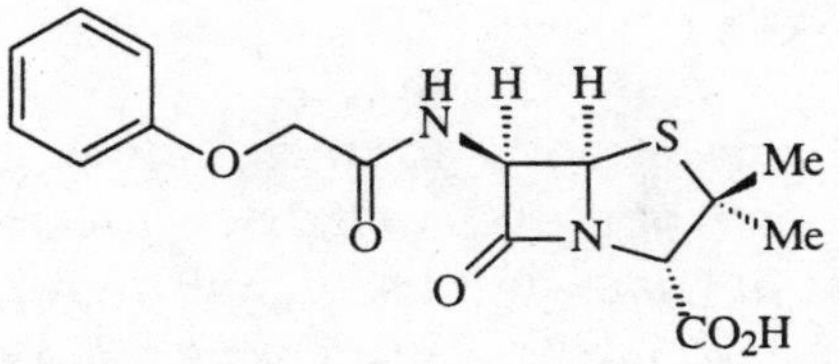

67

Most of the penicillins and cephalosporins in use today are semi-synthetic, in that they are derived from penicillin G or V. Two chemical transformations have been of particular importance in the development of the newer β-lactam antibiotics. The first involves side chain cleavage to produce 6-amino peni-cillanic acid (6-APA) and is shown in Figure 1.3, while the second involves ring expansion chemistry developed by Eli Lilly in 1968 and shown in Figure 1.2. This method allowed the synthesis of important cephalosporins like cephalexin (**68**).

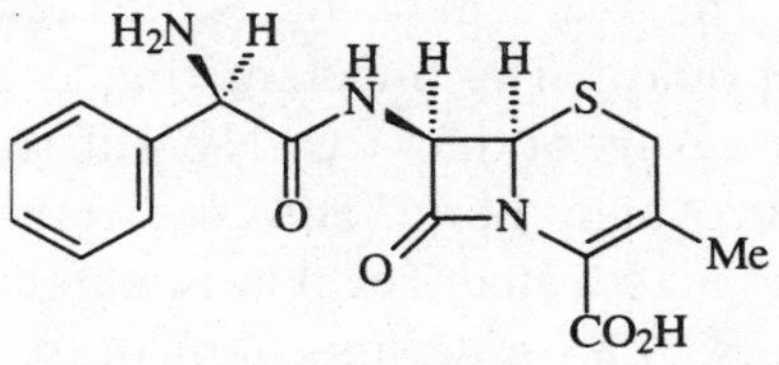

68

In the future, new and more effective methods of entirely chemical syn-thesis could be made available by understanding the biosynthetic routes. The availability of pure, cloned, biosynthetic enzymes will facilitate the cloning of the corresponding genes. This in turn will reveal the complete amino acid sequences of the enzymes and thus bring nearer the determination by X-ray analysis of their three-dimensional structures, at high resolution, if suitable crystals can be obtained. Also, the cloning of genes might also open the way to the construction of higher yielding organisms or of organisms containing new combinations of genes that endowed them with new biosynthetic abilities. Cloning of the genes that encode the isopenicillin N synthetase and the deacetoxycephalosporin C synthetase/hydroxylase in *C.acremonium* is now accomplished. It may well be that the cloning of this and the other genes that code for the series of enzymes in the biosynthetic pathways to penicillins and cephalosporins will lead to major advances in this field during the next decade.

Mode of action of β-lactams

The cell walls of bacteria are essential for their normal growth and development. The major component of the cell wall that confers rigidity is the heteropolymer peptidoglycan. In Gram-positive organisms, the cell wall is 50–100 molecules thick, but only 1–2 molecules thick in Gram-negative bacteria. The peptidoglycan is composed of linear strands of two alternating aminosugars—*N*-acetyglucosamine and *N*-acetylmuramic acid—that are cross-linked by peptide chains (see Figures 2.5 and 2.6).

The composition of the peptide cross-links is characteristic of individual microbial species. The completion of the cross-link is accomplished by a transpeptidation reaction that occurs outside the cell membrane, while the enzyme itself is membrane-bound. In the enzyme reaction, the terminal glycine residue of the pentaglycine bridge (see Figure 2.5) is linked to the fourth residue of the pentapeptide (D-alanine), releasing the fifth residue, another D-alanine. It is this step that is inhibited by the β-lactam antibiotics. The conformation of penicillin was found to be very close to that of D-alanyl-D-alanine (see Figure 4.3). The β-lactam occupies the D-alanyl-D-alanine substrate site of the transpeptidase. The β-lactam ring is under considerable strain, and it is broken by cleavage at the —CO—N bond, to leave the antibiotic linked to the enzyme by a covalent bond. Thus the transpeptidase has been irreversibly acylated, and so inactivated. The β-lactamases which destroy penicillins inactivate penicillins by a similar mechanism (see Figure 5.1). These common modes of reaction are unsurprising given the angle strain possessed by the fused β-lactam ring. The orbital of the nitrogen atom that contains the lone pair of electrons cannot easily overlap with the π-system of the adjacent carbonyl group (as is the case in acyclic amides). Stabilization through electron delocalization is thus impossible. The carbonyl in consequence becomes more susceptible to nucleophilic attack.

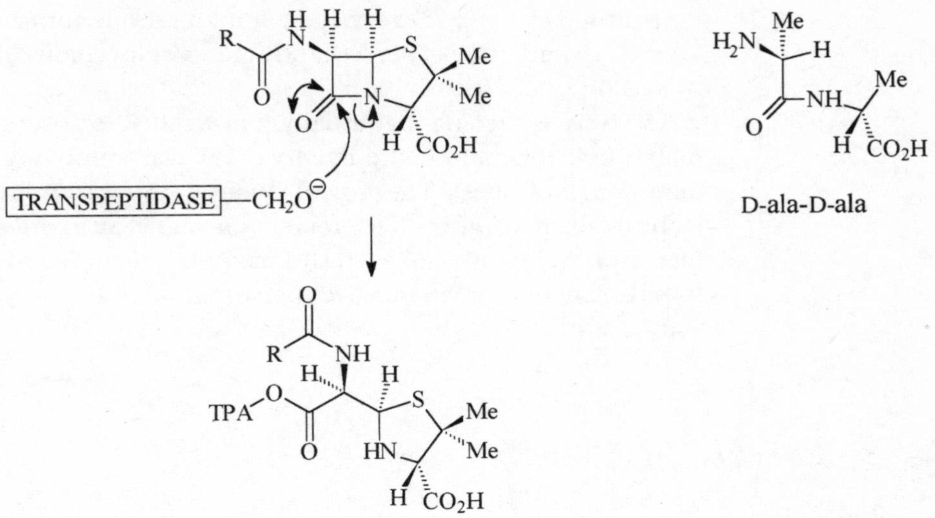

Figure 4.3

D-ala-D-ala

The β-lactam ring is also susceptible to attack by the side-chain carbonyl (as shown in Figure 4.4). The greater stability of penicillins with electron-with drawing groups on the carbon alpha to the carbonyl can be ascribed to the lower nucleophilicity of the carbonyl oxygen in these compounds.

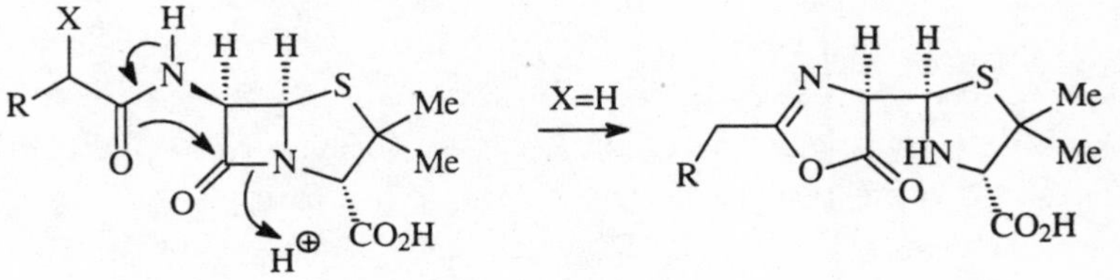

Figure 4.4

[X=NH$_2$, Cl, heterocyclic ring: reaction is less favourable]

The β-lactam antibiotics also exert their antibacterial effect by inactivating high molecular weight penicillin-binding proteins (PBPs). All bacteria have such proteins; for example *Staphylococcus aureus* has four PBPs, while *Escherichia coli* has at least seven. The high molecular weight PBPs of *E. coli* are believed to possess an amino terminal peptidoglycan transglycosidase domain and a carboxy terminal penicillin-sensitive transpeptidase domain. These enzymes are inserted into the membrane only at their amino termini. The PBPs vary in their affinity for β-lactams, although the links always become covalent. Apart from the transpeptidases, PBPs include proteins responsible for maintenance of rodlike shapes, and for septum formation at cell division. There is a similarity between the structures of low molecular weight PBPs and Class 1 β-lactamases.

The lysis of bacteria that usually follows their exposure to β-lactams is ultimately dependent upon the activity of cell wall autolytic enzymes, autolysins, or murein hydrolases. The growth of autolysin-resistant bacteria can be arrested by β-lactams, but the bacteria remain viable. Such organisms are said to be 'penicillin-tolerant', and such strains of streptococci and staphylococci have been isolated from patients with persistent infections.

5 Bacterial resistance to antibiotics

Introduction

As early as 1940, Chain and Abraham observed that some bacteria were insensitive to penicillin, whilst others possessed mechanisms for deactivating their crude antibacterial preparations. The number of resistant organisms increased sharply once the penicillins were in widespread use, since the bacteria were forced to develop new mechanisms to combat the effects of these agents in order to survive and proliferate; their survival as a species depended upon it. The situation was exacerbated by the large-scale use of the penicillins in animal feeds during the 1950s. This ensured that strains of bacteria encountered the antibiotics years before they might otherwise have done so.

The staphylococci have always been in the vanguard where resistance is concerned and as each new antibiotic was introduced, so strains of resistant staphylococci have emerged. In the early 1960s about 10% of strains were resistant to penicillin G. The figure now is close to 100% for most of the simpler penicillins. Some organisms such as the gonococci and pneumococci were initially slow to acquire resistance mechanisms, but during the last two decades there has been a sudden and alarming increase in their resistance to numerous types of antibiotics.

Resistance has always appeared first in hospitals, where patients with life-threatening conditions are treated with the newest and most sophisticated antibiotics. But more recently there has been a growing concern at the prevalence of resistant strains in day centres and residential homes for the elderly. Many of the day visitors and residents receive maintenance doses or frequent doses of antibiotics, and the development and spread of resistant strains is thus guaranteed. These two 'ecosystems' are not isolated from each other or indeed from the rest of the world, and the insidious spread of resistant strains is inevitable.

The first question that must be addressed is, 'How to bacteria become resistant?' To understand this it is necessary to consider the biochemical mechanisms behind resistance.

The biochemical basis of resistance

Resistance mechanisms (based upon biochemical changes) can be classified into three categories:

- modification of the target site or enzyme
- prevention of access for the antibiotics
- production of enzymes that destroy or inactivate the antibiotic

As seen in Chapter 3, most antibiotics inhibit one or more bacterial enzymes, and mutation of the genes producing these enzymes can lead to the production of modified enzymes which have a greater affinity for their natural substrate(s) and/or a lower affinity for the antibiotics. This is typically the way bacteria have developed resistance to the sulfonamides. They produce a modified form of dihydropteroic acid synthase which has a lower affinity for sulfonamides than for p-aminobenzoic acid—the natural substrate.

In *Escherichia coli* a different mechanism operates, and mutation of the gene for the enzyme dihydrofolate reductase leads to the production of both normal enzyme and modified enzyme that has an affinity for trimethoprim 20,000 times lower than the native enzyme. This second type of enzyme can thus catalyse the production of the folic acid required for bacterial growth.

The existence of resistance in *E. coli* and other enterobacteria is a major problem since these represent an endogenous reservoir of resistant organisms always available for the transfer or assimilation of resistant genes. As many as 30% of *E. coli* strains found in the community, and up to 50% of strains found in hospitals are resistant to the broad spectrum antibiotic amoxicillin.

In addition to enzyme modification, other structural changes can lead to inefficient uptake or binding of antibiotics. For example, new or altered bacterial genes code for structurally modified penicillin binding proteins, or cause changes in the permeability of the bacterial membrane to certain antibiotics, or may even initiate mechanisms for the active expulsion of antibiotics.

The worrying increase in the prevalence and severity of *Streptococcus pneumoniae* infections is primarily due to the production of altered forms of penicillin-binding proteins. These bacteria are the most common cause of pneumonia and meningitis. In 1941, 10,000 units of penicillin G administered four times per day for 4 days cured patients with pneumonia caused by this organism. Today, 24 million units of one of the newer penicillins might be effective, but there is no guarantee of success.

Resistance mechanisms used by this organism also include modification of the target of antibiotic attack. Production of a methylase enzyme that methylates a crucial adenine residue of the 23S ribosomal RNA prevents binding of macrolide antibiotics such as erythromycin A (see Figure 6.9). This antibiotic is widely used in France, and about 25% of strains of *P. pneumoniae* in France are resistant to this organism, whereas there is little resistance in the USA where the drug is unpopular because of its gastro-intestinal side effects.

With strains of streptococci, enterococci and camplylobacter, resistance is achieved through the production of a protein that associates with the ribosome and alters it such that antibiotics can no longer bind effectively.

Finally, bacterial strains may simply be insensitive to the antibiotics because they lack a structural element that is normally destroyed by the antibiotic. A prime example of this phenomenon is provided by the Pseudomonads which have very little peptidoglycan in their cell wall and are thus virtually inert to the effects of penicillins and cephalosporins.

All these mechanisms are essentially passive in that they rely on subtle changes in structure and affinity of enzymes or binding proteins. Most bacteria have also evolved mechanisms for destroying or inactivating antibiotics, and the best studied examples are the beta-lactamases and chloramphenicol acetyl transferases.

β-lactamases

As early as 1944 it was known that strains of *Staphylococcus aureus* produced enzymes, known initially as penicillinases, but more recently as β-lactamases, that would destroy penicillins. As increasing numbers of different penicillins and then cephalosporins were introduced, it became clear that an entire family of β-lactamases was operating and that these were produced by most Gram-negative organisms and by some Gram-positive organisms. Even species that have always been susceptible to penicillins and cephalosporins are now showing clear evidence of resistance through the production of β-lactamases. These include *Haemophilus influenzae*, *Neisseria gonerrhoea*, *Enterococcus faecalis*, *Neisseria meningitidis* and various Salmonella species. All these organisms are responsible for serious life-threatening conditions that were believed to have been eradicated or at least brought under control.

The probable mode of action of these enzymes is shown in Figure 5.1. It is significant that there is a great similarity between the first step of this mechanism and the one involved in the normal mode of action of these antibiotics (see Figure 4.3). With both penicillins and cephalosporins the β-lactam ring is cleaved to produce unstable intermediates which then fragment. Initial attempts to prevent attack by β-lactamases centred on the strategy of providing a bulky 'shield' in the acyl side-chain. Methicillin (**69**) was the first analogue to demonstrate the efficacy of this strategy, though its lack of an electron-withdrawing group alpha to the carbonyl meant that it had limited stability towards acids. The semi-synthetic oxacillin (**70**) possesses both a 'shield' and an electron-withdrawing group. It is both acid-resistant and β-lactamase-resistant.

Figure 5.1

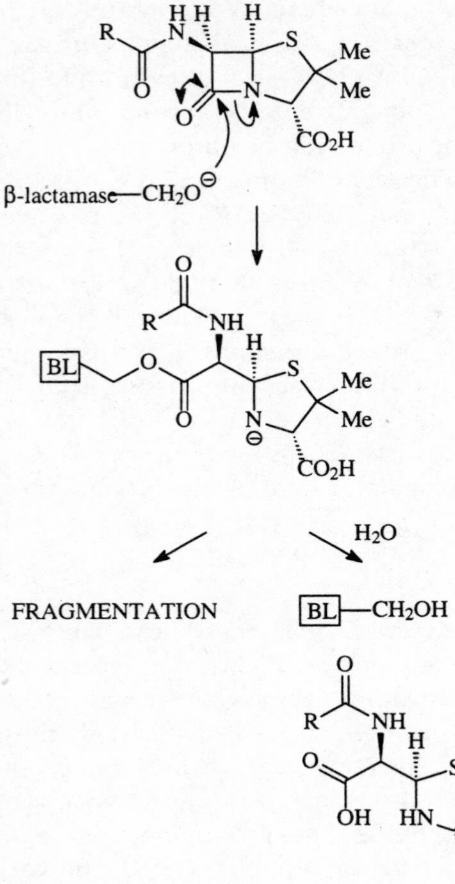

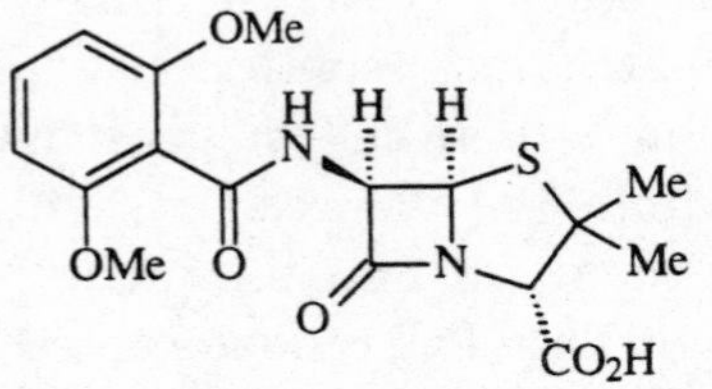

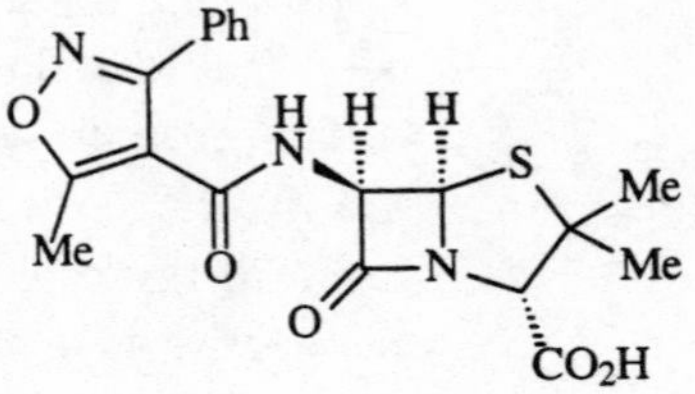

Several classification systems have been devised for the various enzyme types, but the one based upon amino acid and nucleotide sequence studies is perhaps the most informative. Class A enzymes have a serine residue at the active site where acylation occurs, a molecular weight of about 29 kDa and they preferentially hydrolyse penicillins. Class B enzymes are zinc metallo-enzymes and have an active site thiol that must bind to the zinc atom for acylation to occur. Class C enzymes also employ an active site serine for acylation, but have no amino acid sequence homology with Class A enzymes and also have molecular weights of about 39 kDa.

Chloramphenicol acetyl transferase

Chloramphenicol (**25**) is widely used as a broad-spectrum antibiotic in the treatment of typhoid fever, paratyphus, spotted fever, infectious hepatitis, dysentry, malaria and diphtheria. Bacterial resistance to chloramphenicol is normally conferred by the enzyme chloramphenicol acetyl transferase (CAT). The enzyme catalyses the *O*-acetylation of the antibiotic using acetyl coenzyme A (acetyl CoA) as acetyl donor. The acetylated chloramphenicol cannot bind to bacterial ribosomes and so is inactive as an inhibitor of protein synthesis. A number of variants of CAT all show a marked degree of amino acid sequence homology, and it is likely that they are all trimeric enzymes with identical subunits of molecular weight around 25 kDa. The proposed mode of action of the enzyme is shown in Figure 5.2. Chloramphenicol and acetyl CoA approach the active site from opposite ends of a 25Å 'tunnel' to bind in close proximity to an essential histidine residue (his-195). This formation of a ternary complex between enzyme and substrates is in agreement with data obtained from steady-state kinetic experiments. Catalysis proceeds by his-195 acting as a general base, abstracting a proton from the 3-hydroxyl of chloramphenicol, thus promoting nucleophilic attack at the thioester carbonyl group of acetyl CoA. Site-directed mutagenesis studies on the enzyme have provided evidence that the subsequent transition state is stabilized by the hydroxyl group of a nearby serine residue.

Figure 5.2

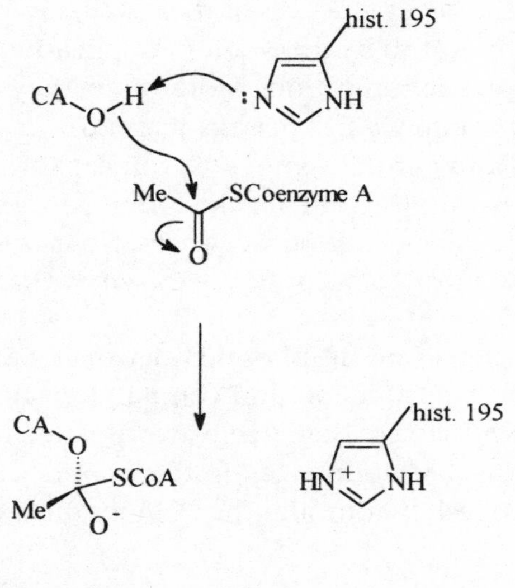

[CA is chloramphenicol]

Other acetyl transferase enzymes are responsible for the acetylation of many of the aminoglycoside antibiotics including streptomycin, spectinomycin and the gentamycins. These antibiotics also suffer from the attentions of various bacterial phosphotransferases and adenyl transferases which phosphorylate and adenylate them. A new enzyme has been discovered recently that has the ability to acetylate and phosphorylate most of the aminoglycosides except streptomycin and neomycin.

The genetic basis of resistance

Bacteria have two mechanisms by which they may become resistant through genetic changes: they may produce new genes through mutation but they may also accept pieces of genetic material from other (resistant) species. This transfer usually takes place via extrachromosomal elements known as plasmids and transposons.

Where the target for the antibiotic is a bacterial enzyme, modification of the shape of the active site can usually be achieved following a change in one or more nucleotide bases of the DNA sequence coding for the enzyme. The resultant change in the amino acid sequence (for example a polar residue to non-polar residue) may result in a subtle change in the shape or binding characteristics of the active site. However, this is an extreme case, and it is usually necessary for several base changes to occur before the required change in shape or function results. It is perhaps for this reason that bacteria usually evolve resistance mechanisms through the production of new genes and their enzyme products rather than through reliance on point mutations. Where analysis of the base sequences of the new genes has been accomplished, they sometimes appear to be the result of duplication of the target gene followed by appropriate modification. More often they are genes introduced from another bacterium via transfer of a plasmid.

Plasmids and chromosomes are both capable of individual existence and replication within the cell, but the former are 50–100 times smaller than the latter and are transferable between cells. Not surprisingly the gene products of plasmids are most commonly the enzymes that inactivate or destroy antibiotics.

The transfer of these plasmids between bacterial cells occurs by two mechanisms: transduction and conjugation. In the first, the plasmid gene passes into the bacterial cell while incorporated into the genetic material of a bacterial virus (bacteriophage). In the second, the gene is transferred during the process of cell mating and the DNA is donated via hair-like projections or sex pili.

Whatever the mode of transfer, these resistance (R) plasmids often carry genes that specify different enzymes required for the destruction of antibiotics. These genetically linked resistance determinants appear to accumulate because certain polynucleotide sequences (containing the discrete genes coding for a particular enzyme) known as transposons can be translocated and inserted into the plasmid without the need for sequence homology with other polynucleotides.

This assemblage of resistance genes on R-plasmids is the most worrying feature of antibiotic resistance. It means that during a bacterial infection, the invading susceptible species can encounter resistant species in the gut and in one encounter assimilate resistance genes that code for enzymes that will destroy a range of antibiotics. This is the reason that strains of bacteria in hospital environments are so dangerous, since they have inevitably encountered most antibiotics and the surviving strains will be resistant to all of them.

For example, clinical isolates of methicillin-resistant staphylococci carry a resistance gene (*mec* A) that confers an intrinsic resistance to all β-lactams. The *mec* A gene is on a 30–40 kb DNA element (*mec*) of unknown origin and nature that integrates into a fragment of the *Staphylococcus aureus* chromosome. The gene codes for an additional low-affinity penicillin-binding protein.

The need for new antibiotics with novel modes of action is thus obvious, and these are the subject of the final chapter in this book.

New antibiotics and new strategies

Introduction

The pharmaceutical industry has responded to the ever-growing problem of antibiotic resistance in two ways: through the design and synthesis of more sophisticated β-lactam antibiotics and via the production of completely novel structures with new modes of activity.

β-Lactam antibiotics

Tens of thousands of semi-synthetic penicillins and cephalosporins have been produced during the past 20 years. Most of these compounds have been produced via acylation of 6-aminopenicillanic acid (**17**) or of 7-amino-cephalosporanic acid (**71**), and they have been designed so as to have improved bioavailability, better penetration through the bacterial cell wall, high affinity for the target enzymes, and good stability against beta-lactamases. The good activity of ampicillin (**18**) was improved through *para*-hydroxylation of the aromatic ring to produce amoxicillin (**19**) in 1971, which had better characteristics with regard to absorption and more rapid onset of antibacterial activity. Acylation of the amino group to produce compounds like piperacillin (**72**) (prepared in 1977) greatly enhanced the spectrum of activity, and these antibiotics typically have activity against Gram-positive cocci, Enterbacteriaceae, *Bacterioides*, *Haemophilus* and *Pseudomonas*. The discovery that the naturally occurring cephamycins (see later) possessed a methoxyl group adjacent to the amide side-chain, encouraged the synthesis of a range of similar penicillins such as temocillin (**73**) (prepared in 1982). This is, like the cephamycins, effective against several organisms producing beta-lactamases, but has little activity against Gram-positive organisms and *Pseudomonas*.

Poor uptake of penicillins by absorption through the gut wall often results if the drugs are too polar, and this problem has been addressed through the synthesis of prodrugs. For example, ampicillin (**18**) is dipolar due to the pres-

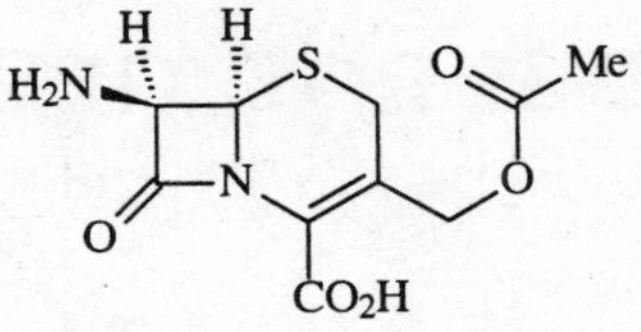

71

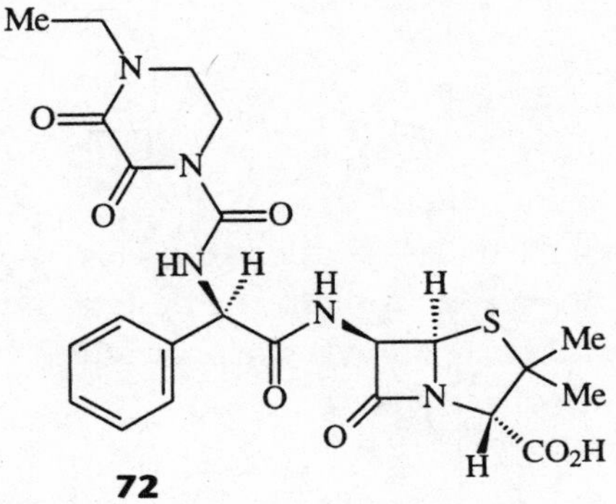

72

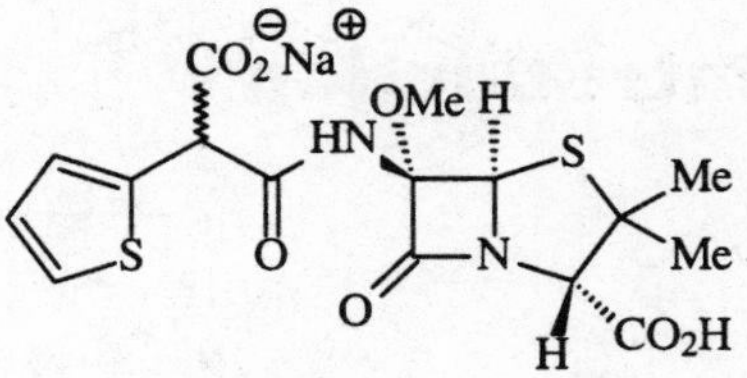

73

ence of the amino and carboxyl functions and is poorly absorbed, while the prodrugs pivampicillin (**20**) and talampicillin (**74**) are much more lipophilic and are readily absorbed. Once in the bloodstream they are hydrolysed to ampicillin by the action of non-specific esterases. Similarly, the phenyl ester carfecillin (**75**) is a prodrug of carbenicillin (**76**). This latter drug is particularly effective against Gram-negative organisms especially *Pseudomonas aeruginosa* which is a particular problem for burns victims.

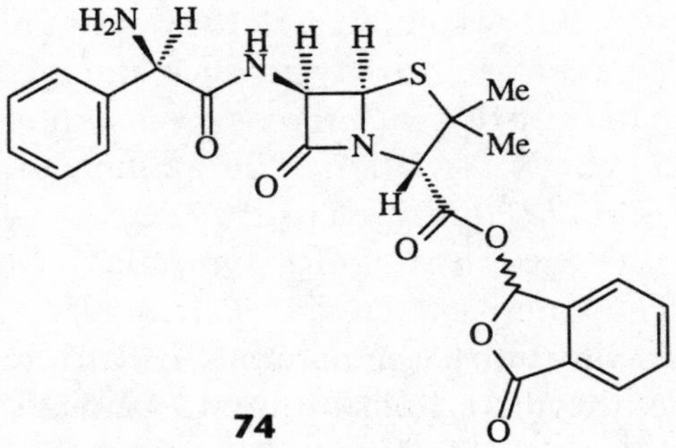

74

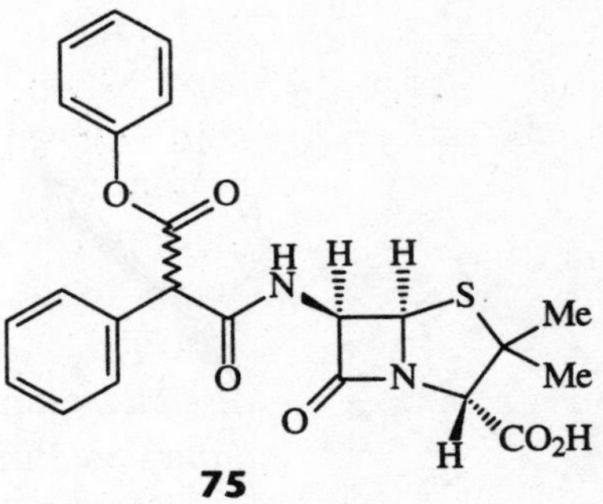

75

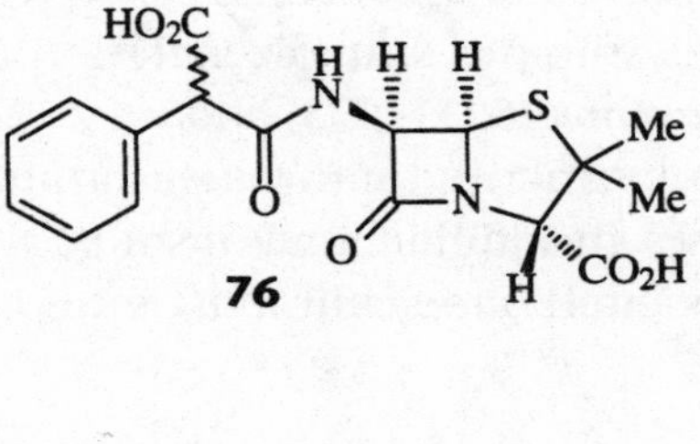

76

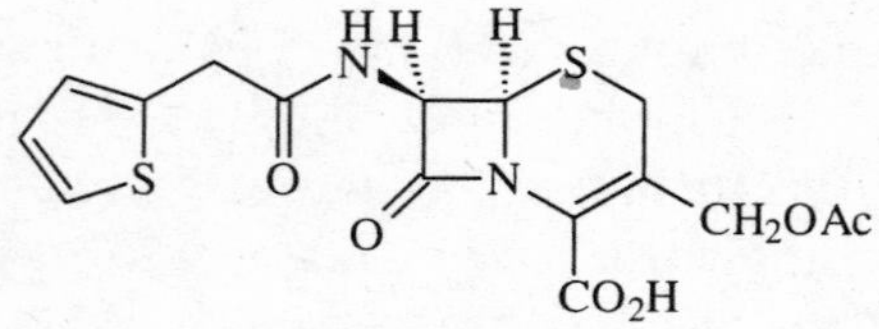

77

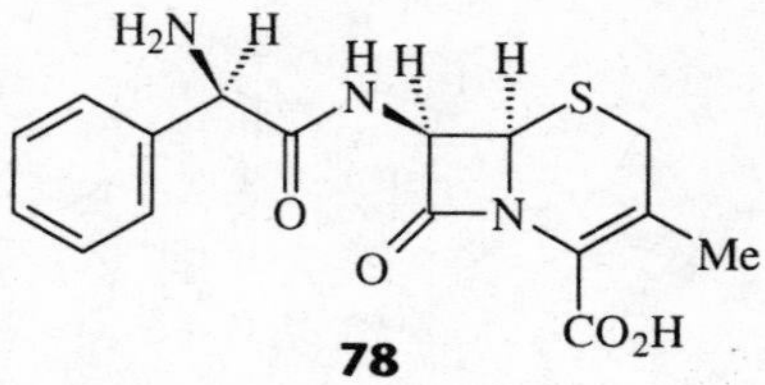

78

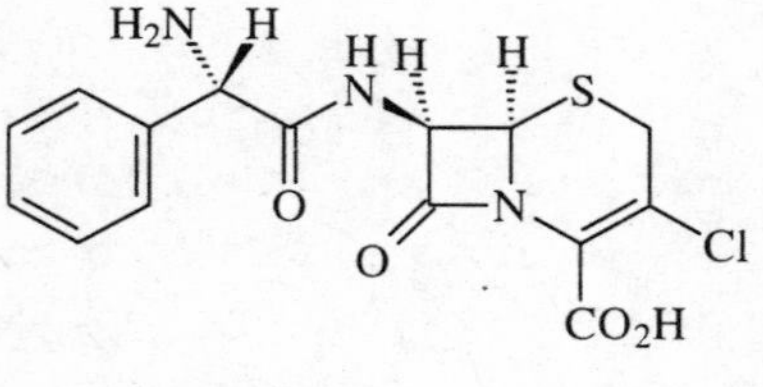

79

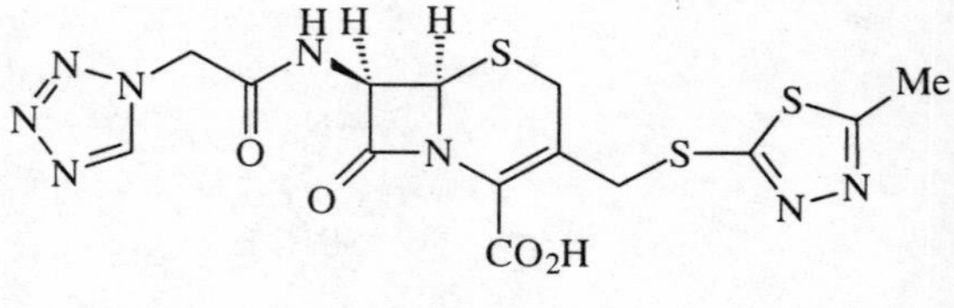

80

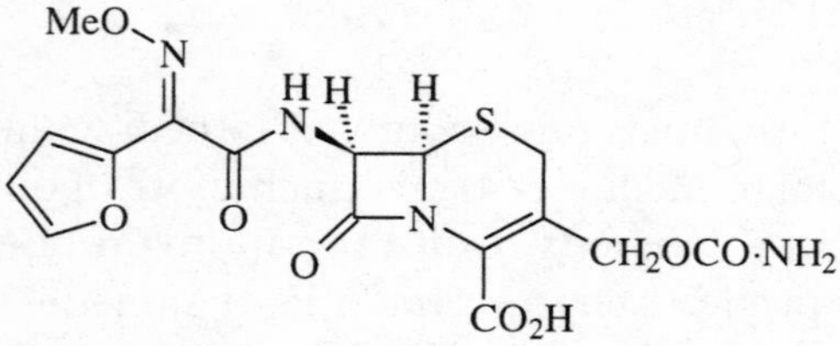

81

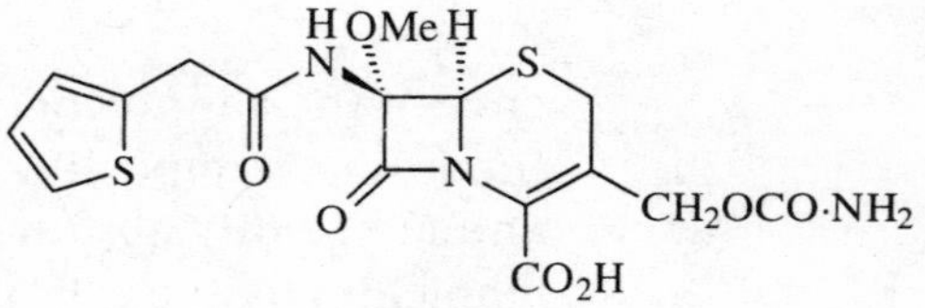

82

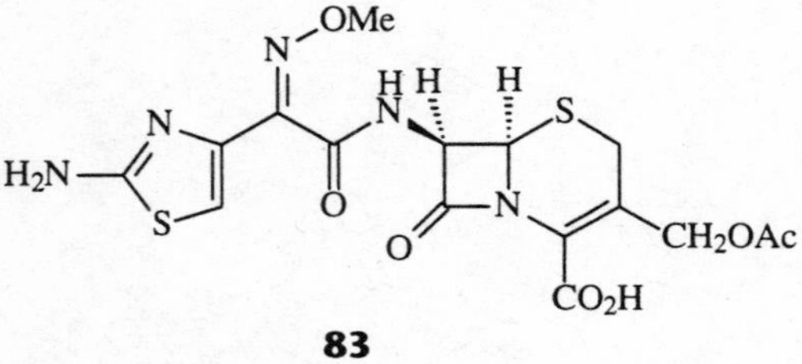

83

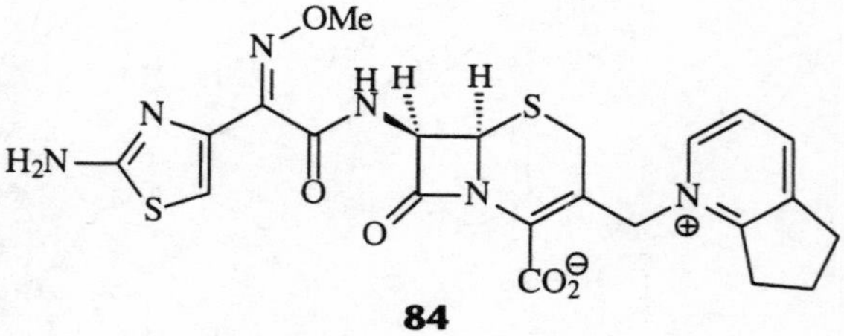

84

The cephalosporin structure has more sites at which structural modification may be made, and the first clinically useful (though not orally active) cephalosporin, cephalothin (**77**) (1962), has evolved to provide analogues of the penicillins, such as cephalexin (**78**) (1967) and cefaclor (**79**) (1974) which are orally active, and thence into the second generation compounds such as cefazolin (**80**) (1970), cefuroxime (**81**) (1975) and cefoxitin (**82**) (1972). Each of these has a good spectrum of activity and (at least initially) a high level of stability towards β-lactamases. In addition, they are more acid stable than penicillins and cause less allergic responses, but most must be injected. Further modification led to the discovery of cefotaxime (**83**) (1975) which has good activity against Gram-positive organisms and outstandingly good activity against most Gram-negative organisms including *Pseudomonas* and multiply resistant strains of *E.coli*. Not surprisingly this led to a massive effort to find even better compounds, and hundreds of analogues of cefotaxime were prepared through a systematic modification of the 3-substituent and the oxime ether group in particular. Resultant compounds such as cefpirome (**84**) combined extreme stability against β-lactamases with an extensive spectrum of activity against Enterobacteriaceae, Staphylococci and *Pseudomonas*. With these aminothiazole antibiotics, the introduction of a methoxyl group at C-7 led to compounds with a poorer profile of activity and stability.

More recently, research efforts have concentrated on improving the oral availability of the cephalosporins, and their resistance to beta-lactamases, since the first generation compounds cephalexin and cefaclor were relatively insoluble and had little resistance to beta-lactamases. Some of the best compounds presently available, such as FK 037 (**85**) and FK 312 (**86**), have a more exotic heterocyclic group at C-3 than do the parent compounds.

85

86

It will be apparent from the variety of structural types involved that a number of key chemical transformations are required for the synthesis of semi-synthetic cephalosporins. Principal amongst these are the displacement of the acetate of 7-ACA (**71**) by a nucleophilic species, e.g. 5-methyl-1-thia-3, 4-diazole-2-thiol for the production of cefazolin (**80**), and introduction of the requisite acyl side-chain. The synthesis of ceftazidime (**87**), which has excellent activity against Gram-negative organisms, exemplifies the usual strategies (Figure 6.1). The complex side-chain moiety is joined to 7-ACA using dicyclohexylcarbodiimide (DCC) as condensing agent. After selective deprotection of the side-chain moiety with trifluoroacetic acid (TFA), the pyridine group is introduced by displacement of the acetate (probably via initial displacement of this group by iodide).

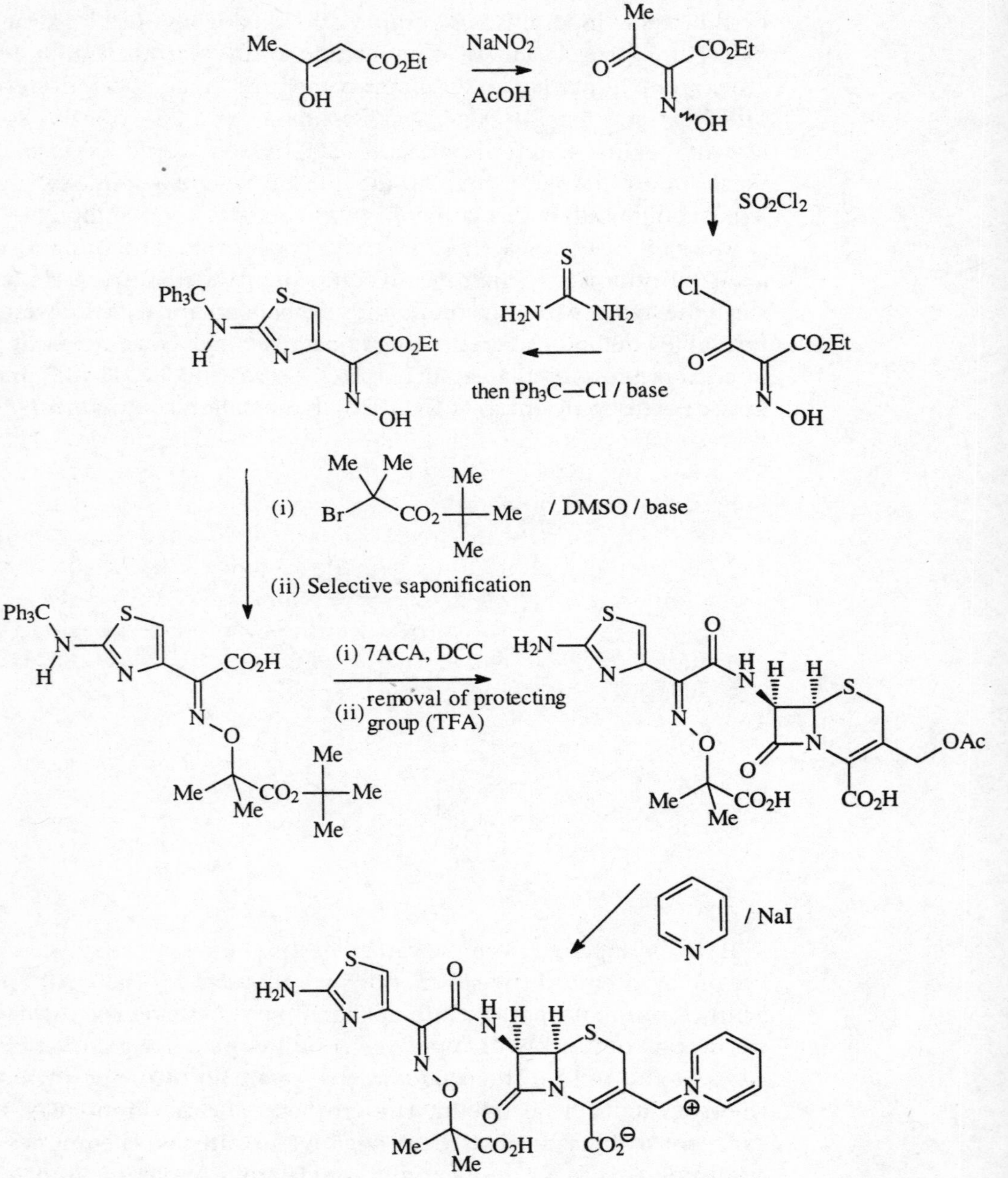

Figure 6.1

87

The 7-methoxy-cephalosporins (cephamycins) based upon the structure of the naturally occurring compound cephamycin C (**88**) from *Streptomyces clavuligerus* have a good spectrum of activities and good stability to β-lactamases. Semi-synthetic analogues like cefoxitine (**82**) are usually made via degradation of the aminoacyl side-chain of (**88**) and then application of the requisite side-chain using similar methods to those shown in Figure 6.1.

The synthesis of the widely used cefaclor (**79**) shows the methods that have been established for its synthesis from a suitably protected penicillin sulph-oxide (Figure 6.2). Reaction with Cl$^+$ (derived from aqueous *N*-chloro-succinimide) and then addition of a Lewis acid like SnCl$_4$ provides the ring-expanded product with an exo-methylene group. Ozonolysis of this yields the enol ether which can be chlorinated. Finally, side-chain removal with PCl$_5$ and reduction of the sulphoxide with PBr$_3$ provides the 7-amino-substrate for application of the 2-aminophenylacetyl side-chain.

Most of the more recent research on beta-lactam antibiotics has centred around the construction of compounds with novel ring systems. 1-Oxa-cephalosporins, which can be prepared from penicillins via the general route shown in Figure 6.3 have a good spectrum of antibacterial activity and excel-lent stability against beta-lactamases. The drug latamoxef (**89**) is a good ex-ample of this class of antibiotic.

The synthesis shown in Figure 6.3 commences with a base-catalysed epimerization of a suitably protected 6-APA sulphoxide. Treatment of the 6-α-aminoacyl product with triphenylphosphine provides an oxazolidine prob-ably via the mechanism shown in the figure. This oxazolidine reacted with chlorine in the presence of bicarbonate to produce a chlorinated oxazolidine. Subsequent hydrolysis after displacement of the chloride by iodide yielded another oxazolidine, and this underwent a stereospecific cyclization in the presence of BF$_3$.etherate to produce the basic skeleton of latamoxef (**89**).

Figure 6.2

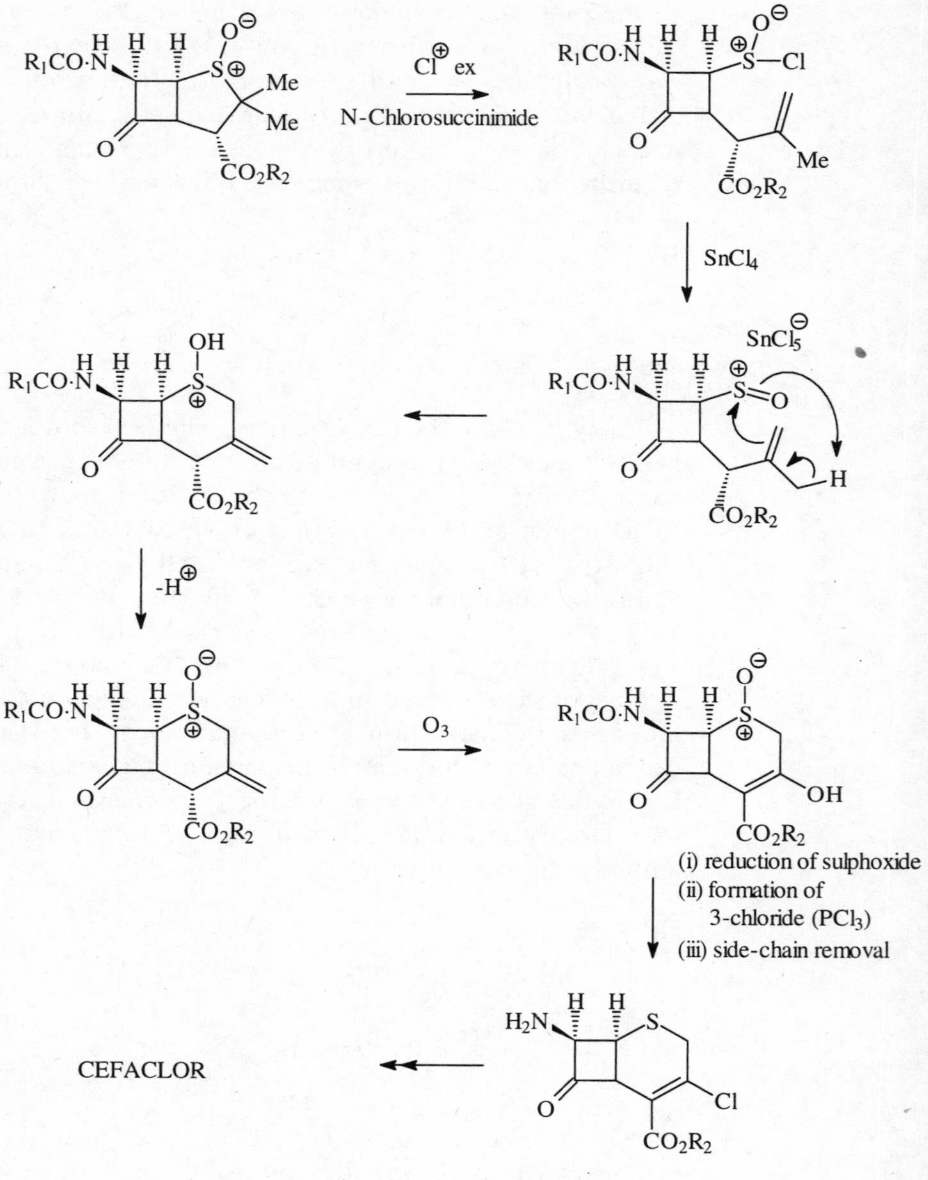

CEFACLOR

However, the greatest success has come from the class of compounds known as carbapenems of general structure (**90**). The first member of this series, thienamycin (**91**), was isolated from *Streptomyces cattleya* in 1976 and proved to have excellent antibacterial activity and resistance to the effects of beta-lactamases. It is active against both Gram-positive and Gram-negative organisms, and against *Pseudomonas* and *Bacteroides* strains. The major drawback is its chemical instability due to the possession of a strained beta-lactam

Figure 6.3

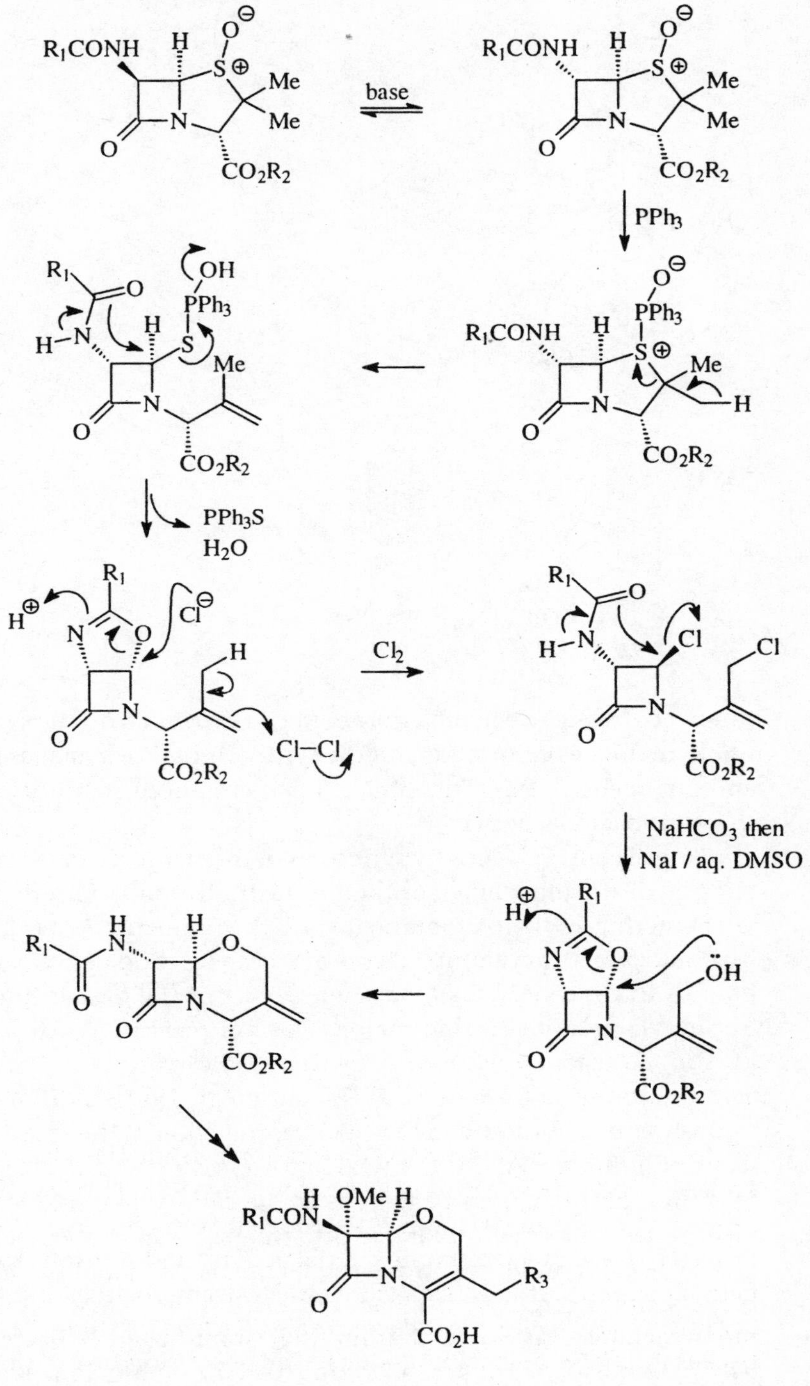

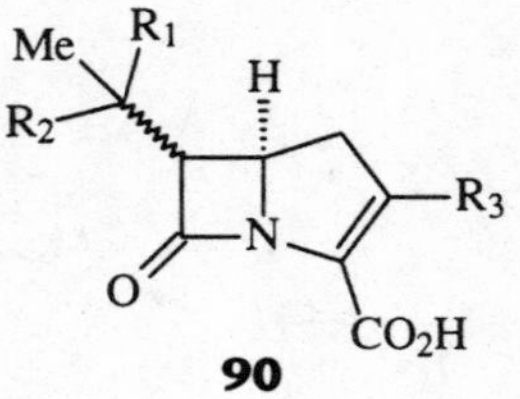

90

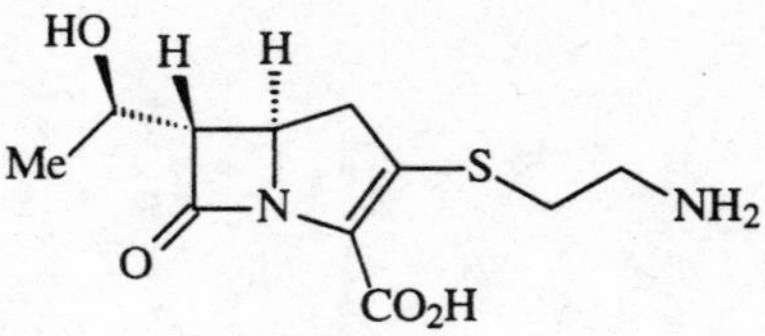

91

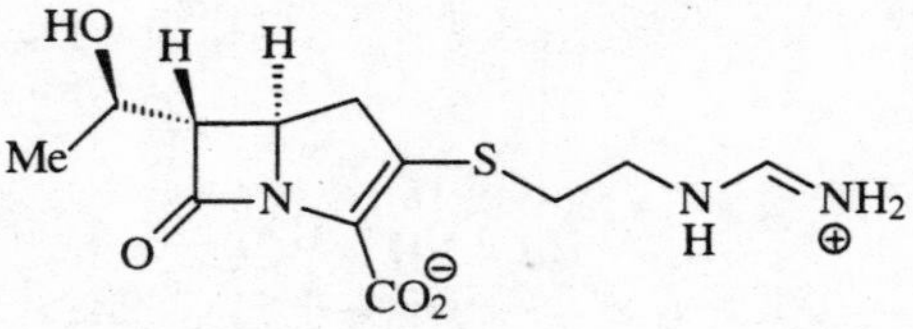

92

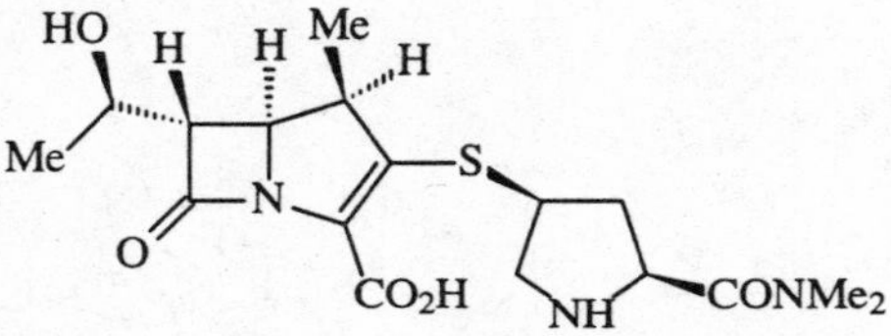

93

ring. In consequence much effort has been expended on the synthesis of more stable analogues culminating in the synthesis of drugs such as imipenem (**92**) and meropenem (**93**). These have much enhanced stability and broad spectrum antibacterial activity.

Many elegant but lengthy syntheses of thienamycin have been carried out, but one of the most efficient (and industrially viable) routes proceeds from (S)(+)-methyl-3-hydroxybutanoate and the imine derived from cinnamaldehyde and *p*-methoxyaniline (Figure 6.4). The reaction between these two compounds in the presence of two equivalents of the base lithium cyclohexyl, isopropylamide yields one major β-lactam product. This participates in a Mitsunobu reaction with Ph$_3$P, diethylazodicarboxylate (DEAD) and formic acid to produce a formate with the anticipated inversion of configuration. Hydrolysis of this ester is followed by reprotection of the secondary alcohol as its *t*-butyldimethylsilyl ether. Oxidative cleavage of the alkene and further oxidation of the resultant aldehyde to the carboxylic acid provided an intermediate that has been elaborated into thienamycin in a variety of ways.

All the compounds mentioned so far are bicyclic, and the first monocyclic beta-lactams were not isolated until 1981. The first of these so-called monobactams was isolated from the Gram-negative bacterium *Chromobacterium violaceum* from a soil sample taken from the Pine Barrens region of

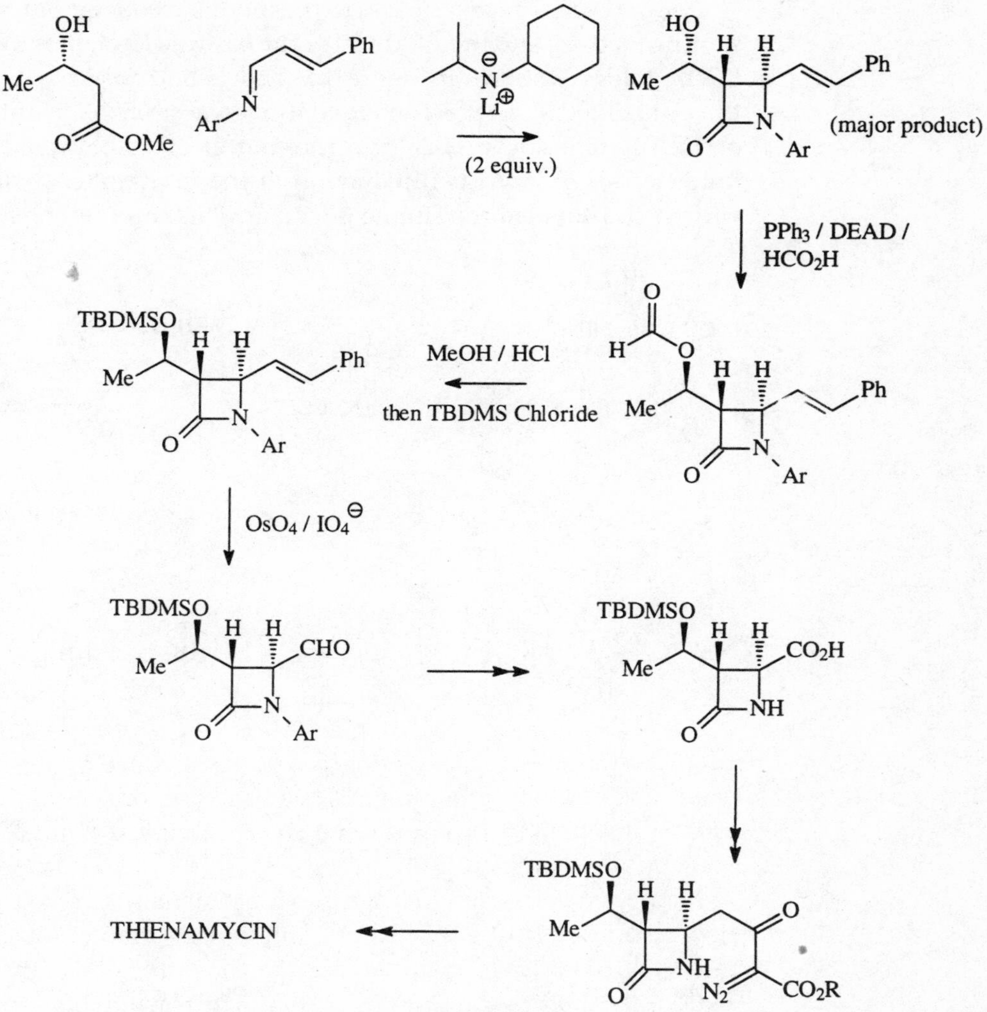

Figure 6.4

southern New Jersey. It proved to have the simple structure (**94**) and a host of other natural and synthetic monobactams have since been isolated or prepared. The synthetic drug aztreonam (**95**) is of particular interest and possesses excellent chemical stability and resistance to beta-lactamases, as well as high potency against Gram-negative bacteria, most importantly strains of *Pseudomonas*.

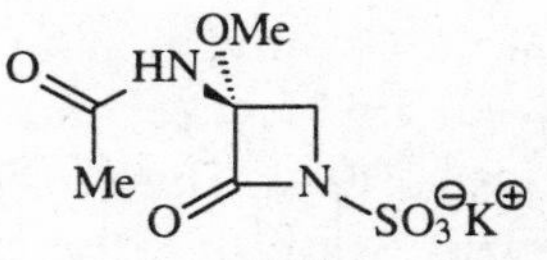

94

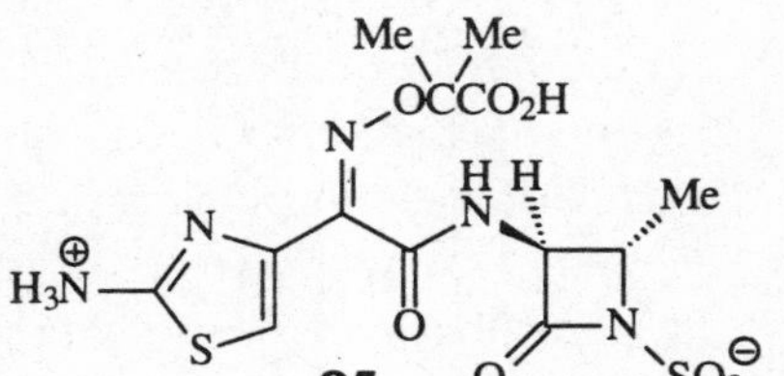

95

One of the best current routes to the monobactams employs (S)-threonine as starting material and the synthesis of the basic skeleton of aztreonam using this methodology is shown in Figure 6.5. The carbobenzyloxy (CBZ) derivative of threonine methyl ester is converted into an acetoxy oxime prior to formation of the β-lactam ring, probably via the chloride. Hydrolysis of the acetate under mild conditions and then reduction of the resultant N-hydroxy compound with $TiCl_3$ provided the simple β-lactam. This can then be reacted with SO_3.

Figure 6.5

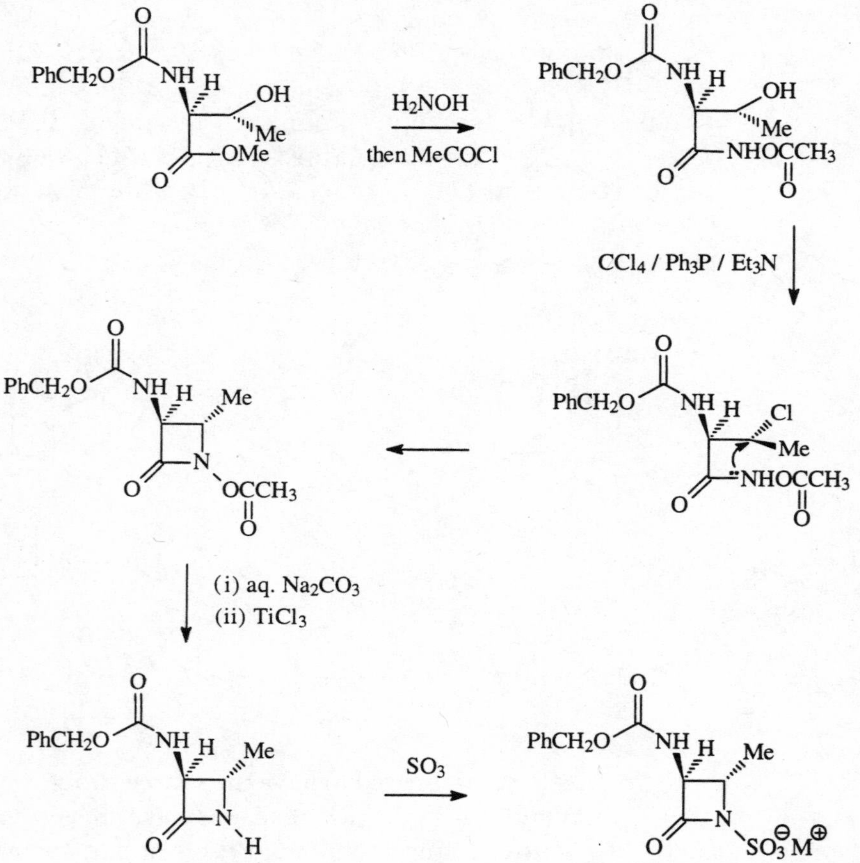

The other major class of monocyclic β-lactams are the nocardicins first isolated from *Nocardia uniformis* in 1975 and exemplified by nocardicin A (**96**). Unfortunately, these compounds show little antibacterial activity.

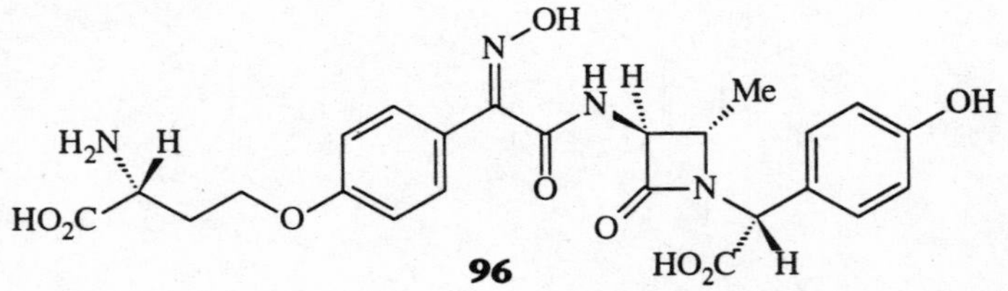

96

Finally, a number of compounds have been isolated that have specific activity against β-lactamase enzymes whilst possessing little intrinsic antibacterial activity. The prime example of this type of compound is clavulanic acid (**97**) from *Streptomyces clavuligerus*. The proposed mode of action of this compound against type 2 β-lactamases (the most frequently encountered) is shown in Figure 6.6 and involves nucleophilic attack of the active site serine of β-lactamases on the β-lactam ring which may further react with subsequent formation of covalently bound enzyme–drug residue products. Alternatively, the initial reaction product hydrolyses to regenerate undamaged enzyme and degradation products from clavulanic acid. A particularly simple synthetic route to clavulanic acid is shown in Figure 6.7. *N*-alkylation is followed by displacement of the thiomethyl group with chloride and subsequent displacement of this by the enol form of the ketone. The double bond stereochemistry can then be inverted by UV irradiation. Clavulanic acid has found considerable clinical utility in combination with broad spectrum antibiotics such as amoxycillin (**19**), a drug combination marketed by Smith Kline Beecham

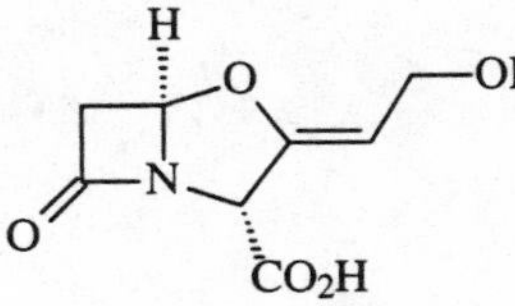

97

Figure 6.6

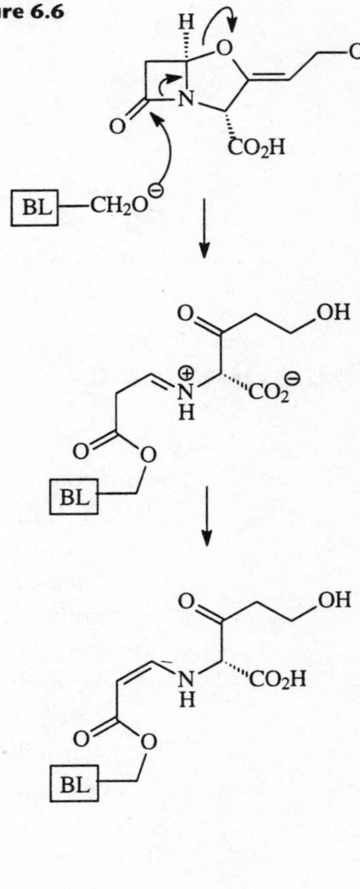

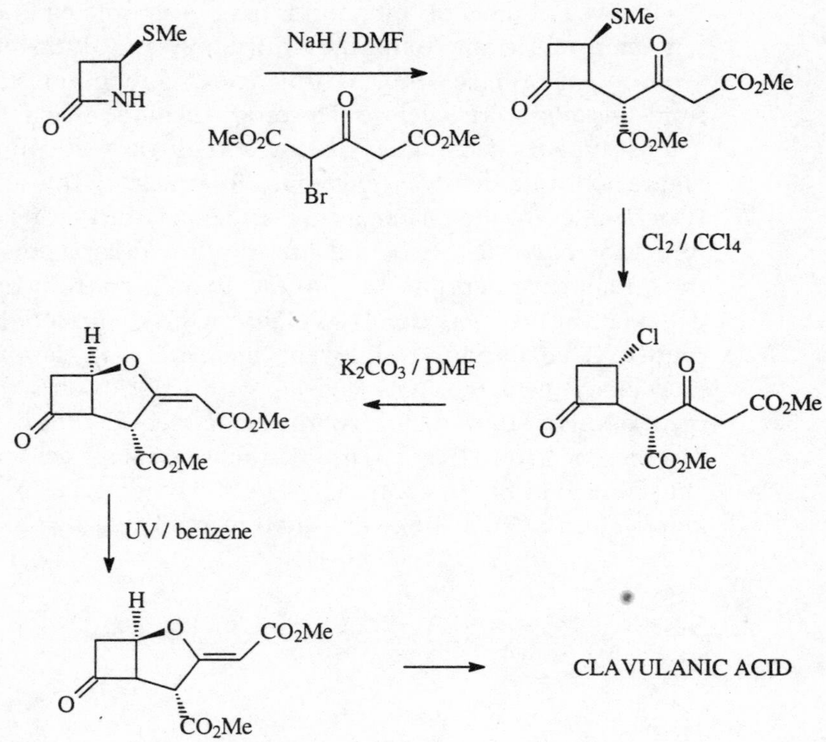

Figure 6.7

under the name Augmentin. The clavulanic acid destroys the β-lactamases thus allowing the broad spectrum penicillin to function without danger of destruction.

The quinolones

The rapid rise in bacterial resistance to the traditional antibiotics such as the β-lactams, the aminoglycosides and tetracyclines has encouraged a continuing search for new classess of compounds with novel modes of antibacterial activity. The quinolones are the most important example of a new generation of antibiotics. The first quinolone, N-methyl-1,4-dihydro-4-oxo-quinoline 3-carboxylic acid (**98**) was prepared in 1949 though its antibacterial properties were not noted until 12 years later. This discovery led to the synthesis of the first commercial quinolone, nalidixic acid (**99**) (1962), but this only exhibited activity against Gram-negative organisms. In addition resistance appeared very quickly and although other compounds such as cinoxacin (**100**) were developed these too had only limited clinical significance.

98

99

100

101

The major breakthrough came in the early 1980s with the synthesis of a range of fluoroquinolones, most importantly ciprofloxacin (**101**), which possesses broad spectrum activity against both Gram-positive and Gram-negative organisms. The synthesis of this drug is shown in Figure 6.8. These fluoroquinolones (and the earlier quinolones) all possess a novel mode of antibacterial activity. They inhibit one particular bacterial enzyme, the so-called DNA gyrase. This controls supercoiling of the chromosome which precedes replication and transcription of the DNA. Although it is now known that the quinolones bind to the N-terminal portion of the A subunit of DNA gyrases, the actual mode of inhibition is unknown. The end result is, however, a failure in replication, transcription and ultimately new protein production and lysis of the bacterial cell.

The bacterial DNA gyrases are related to the mammalian topoisomerases, which control the partial unwinding of chromosomal DNA and the breakage and reformation of DNA strands. Not surprisingly some quinolones possess cytotoxicity towards mammalian cells, though of the hundreds of compounds that have been prepared in recent years, this property has only been significant in a few analogues. The quinolones thus remain one of the major successes of the 1980s though this primacy may be short-lived. Recently,

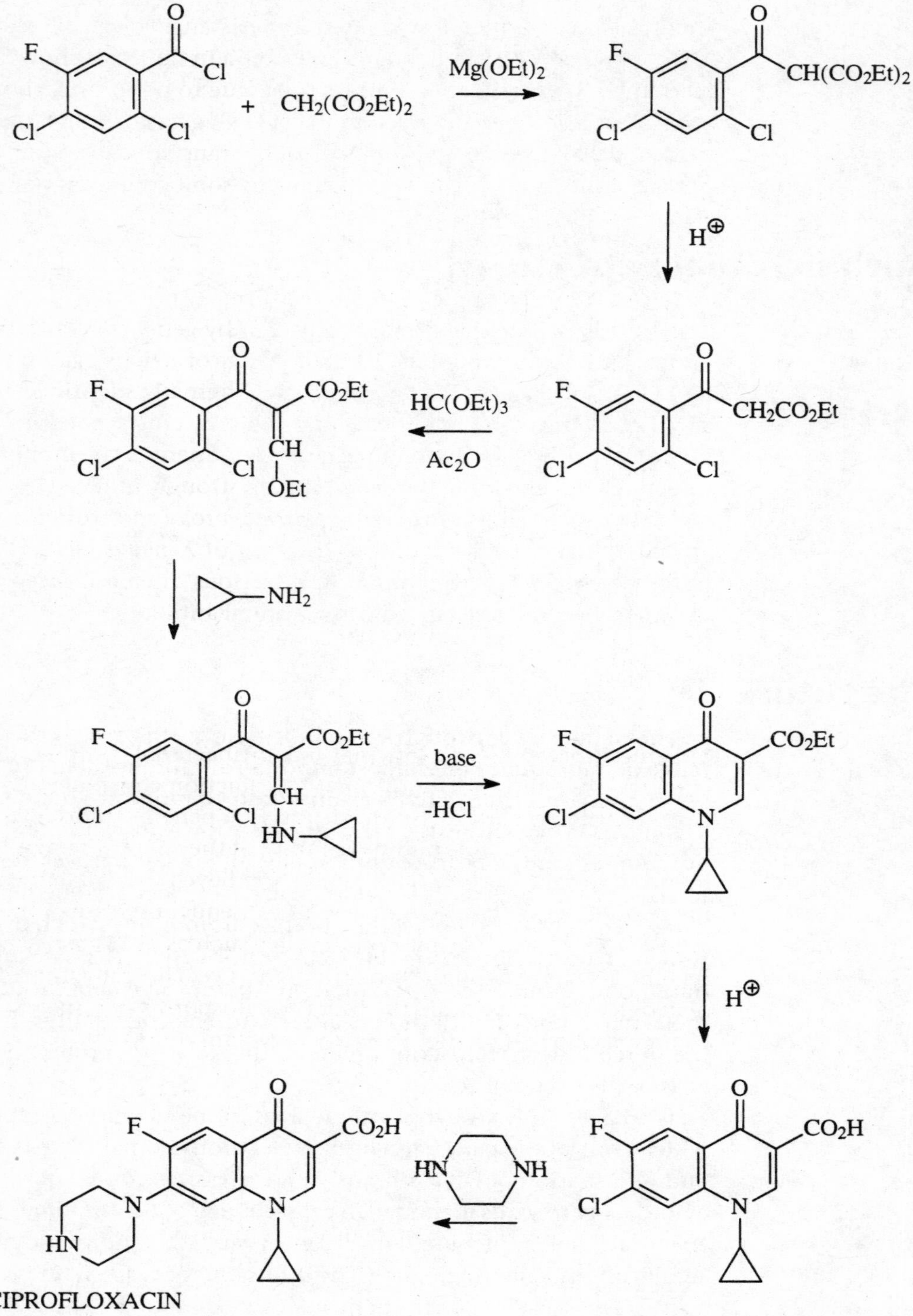

Figure 6.8

CIPROFLOXACIN

there has been a rapid development of resistance to ciprofloxacin with up to 80% of multiply resistant strains of bacteria in hospitals showing evidence of resistance. This resistance appears to be due to point mutations on the bacterial DNA leading to modified forms of DNA gyrase, though there is as yet no evidence that quinolone resistance can be transferred via plasmids. Decreased uptake of the drugs and increased efflux by some strains may also be involved.

Antimicrobial peptides

These are small peptides directly encoded by genes (in contrast to the better known antibiotics) which show a broad range of activity against Gram-negative and Gram-positive bacteria. They were originally identified in insects, but are also produced by mammals, amphibians, crustaceans and plants. Antimicrobial peptide forms an important component of the immune response in insects. Genes encoding the cecropins are strongly induced by various microbial substances. Frogs produce magainins, broad spectrum membrane-active antimicrobial peptides found in the skin of *Xenopus laevis*. These peptides have potential in the treatment of infectious diseases caused by pathogens that have become resistant to conventional antibiotics.

The future

The quest for yet more exotic natural antibacterial compounds and for new chemical structures with novel modes of action continues at a breathtaking pace. The recent synthesis of the dual action penem–quinolones such as (**102**) provide a good example of the ingenuity of the chemists and pharmacologists involved in this research. The penem is believed to acylate the active site serine residue of the transpeptidase with subsequent release of the quinolone to attack DNA gyrase. The pyrazolidinones such as (**103**) recently described by Eli Lilly also have an interesting (though as yet clinically unproven) spectrum of activity. They resemble the best cephalosporins in structure but possess a gamma-lactam ring rather than a beta-lactam ring.

102

103

But perhaps the area where the most exciting discoveries will be made in the next few years is at the genetic level. The studies presently underway concerning the biosynthesis of the antibiotic erythromycin A (**104**) (mentioned in Chapter 5) in *Saccharapolyspora erythrea* provide a good example of these developments. The three large genes that code for the multifunctional enzymes that control biosynthesis have been identified (Figure 6.9), and these enzymes have been subjected to sequence analysis. The overall pathway (Figure 6.10) involves the sequential addition of methylmalonyl SCoenzyme A units to a starter unit of methylmalonyl S-acyl carrier protein. The various enzymes control the level of reduction, dehydration and further reduction of these units to provide ultimately 6-deoxy-erythronolide (**105**), the precursor of erythromycin A (**104**). The abbreviations refer to the individual enzymes which are constituents of the multienzme complex: AT-acyltransacylase; ACP-acyl carrier protein which acts as an anchor for the growing chain; KS-ketosynthase; KR-β-ketoacyl reductase which catalyses the reduction of ketone to alcohol; DH-dehydratase which catalyses dehydration

Figure 6.9

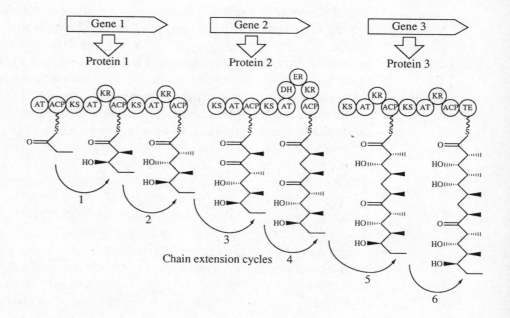

Chain extension cycles

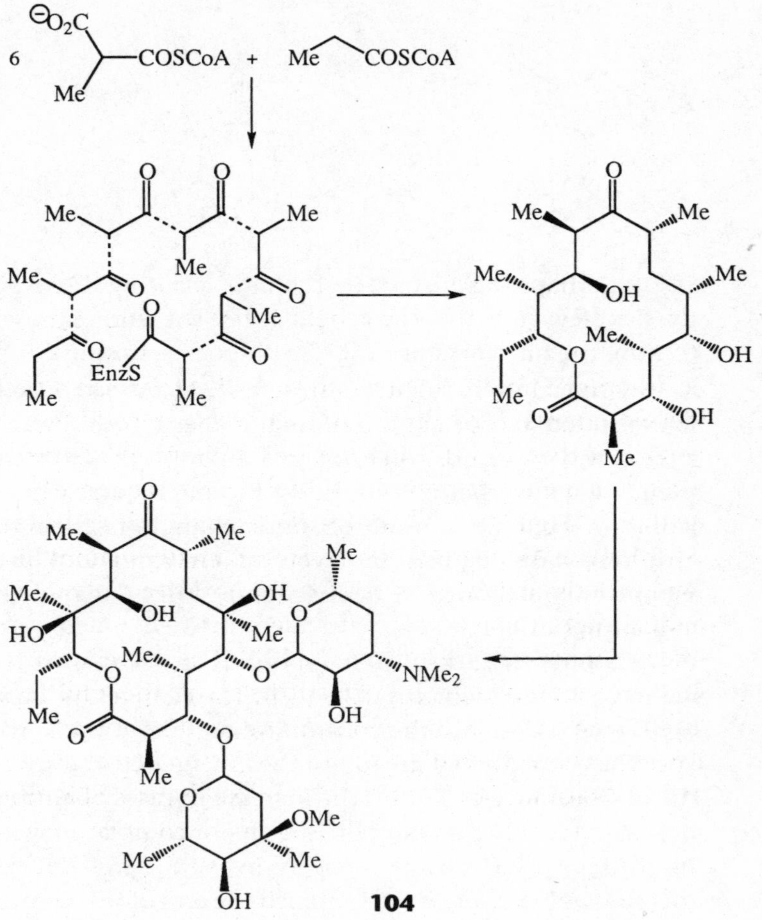

Figure 6.10

to yield *trans*-alkene; ER-enoylreductase which catalyses the reduction of the alkene; and TE-thioesterase which catalyses the cyclisation and associated release from the multienzyme complex.

The implications of this intimate knowledge of the pathway of biosynthesis are that it should be possible to make subtle changes to the genes and thus increase or decrease the levels of the various enzymes. This should result in the formation of novel analogues of erythronolide (and thus erythromycin) which have different numbers of alkenes, hydroxyl and ketone functional groups. Recent work by the groups of Khosla and Leadlay have already demonstrated that such 'genetic engineering' is feasible. For example, using the organism *Saccharopolyspora erythaea*, they showed that if the gene domain coding for the terminal enzyme (that catalyses cyclisation and cleavage from the multienzyme complex) was relocated to the end of the first large gene, the resultant mutant strain produced the novel triketide lactone (**106**). This is clearly produced because the product of the first gene (**107**) (see Figure 6.9)

Figure 6.11

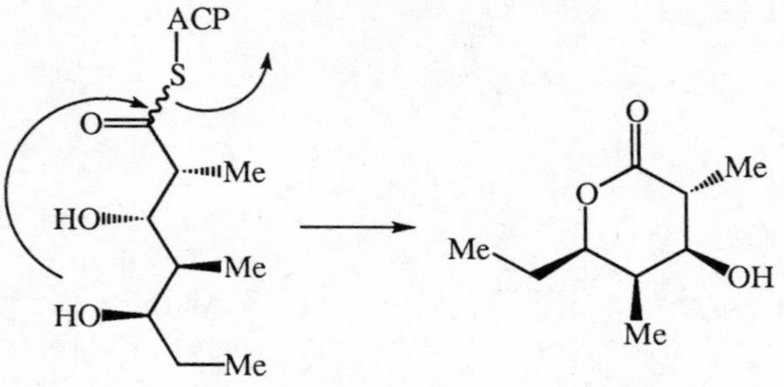

107 → 106

cyclises prematurely as shown in Figure 6.11 instead of acting as a substrate to be transfered to the next multifunctional enzyme. The antibiotic acitivites of these new compounds will be of great interest. This strategy is of course exactly analogous to that employed by the microorganisms in the wild, where evolution occurs as a result of subtle changes in their genetic composition with resultant changes in levels of enzymes and production of novel secondary metabolites. Chemists and molecular biologists will simply be emulating Nature.

The ability of these new compounds to act against the present resistant strains of bacteria remains to be seen. The changes in sexual habits during the 1960s and 1970s, and the continuing presence of poverty in the developing world has contributed greatly to the emergence of drug-resistant organisms. The reappearance of TB and the increase in cases of antibiotic-resistant pneumonia and meningitis leave little room for complacency in the search for new drugs, if we are continue to enjoy lives that are relatively free of bacterial infections. No doubt, as we approach the end of this century and millennium, Man will be as successful as he has been since the observations of Pasteur in the last century, in finding or creating new antibiotics.

Further reading

Chapter 1

An excellent text on the discovery of all classes of antibacterial drugs: Sneader, W. (1985) *Drug Discovery: The Evolution of Modern Medicines*. Wiley.

A fascinating account of the diverse and often unexpected activities of microorganisms is provided by Dixon, B. (1994) *Power Unseen – how microbes rule the world*. W.H. Freeman/Spektrum.

Gales, E.F., Cundliffe, E., Reynolds, P.E., Richmond, M.H. and Waring, M.J. (1981) *The Molecular Basis of Antibiotic Action*, 2nd edn. Wiley.

Franklin, T.J. and Snow, G.A. (1985) *Biochemistry of Antimicrobial Action*, 4th edn. London, Chapman & Hall.

Russell, A.D. and Chopra, I. (1990) *Understanding Antimicrobial Action and Resistance*. Chichester, Ellis Horwood.

Greenwood, D. (ed.) (1995) *Antimicrobial Chemotherapy*. Oxford University Press.

MacFarlane, R.G. (1985) *Alexander Fleming: The Man and the Myth*. Oxford University Press.

MacFarlane, R.G. (1979) *Howard Florey: The Making of a Great Scientist*. Oxford University Press.

Craig, P. (1979) Penicillin: the first half-century. *Chemistry in Britain*, 392.

Chapter 2

Goss, T., Faull, J., Ketteridge, S. & Springham, D. (1995). *Introductory Microbiology*. Chapman & Hall.

Stanier, R.Y., Doudoroff, M. and Adelberg, E.A. (1970) *General Microbiology*. New Jersey, USA, Macmillan.

Davis, B.D., Dulbecco, R., Eisen, H.N. and Ginsberg, H.S. (1990) *Microbiology*, 4th edn. Philadelphia, J.B. Lippincott Co.

Chapter 3

Davis, B.D., Dulbecco, R., Eisen, H.N. and Ginsberg, H.S. (1990) *Microbiology*, 4th edn. Philadelphia, J.B. Lippincott Co.

Atlas, R.M. (1995) *Principles of Microbiology*. Mosby: St. Louis.

Baron, E.J. & Finegold, S.M. (1990) *Diagnostic Microbiology*. Mosby: St. Louis.

Shaw, W.V. (1983) *Chloramphenicol acetyltransferase*. C.R.C. Crit. Rev. Biochem. 14, 1–46.

Sykes, R.B. & Mathew, M. (1976) *J. Antimicrobial Chemotherapy* 2, 115–157.

Chapter 4

Spratt, B.G. and Cromie, K.D. (1988) Penicillin-binding proteins of gram-negative bacteria. *Reviews of Infectious Diseases*, 10, 699.

Aharonowitz, Y., Cohen, G. and Martin, J.F. (1992) Penicillin and cephalosporin biosynthetic genes. *Annual Review of Microbiology*, 46, 461.

Baldwin, J.E., Byford, M.F., Field, R.A., Shiau, C.-Y., Sobey, W.J. and Schofield, C.J. (1993) *Tetrahedron* 49, 3221–3226.

Baldwin, J.E. and Bradley, M. (1990) *Isopenicillin N synthase: mechanistic studies. Chemical Reviews*, 90, 1079.

Roach, P.L., Clifton, I.J., Fülöp, V., Harlos, K., Barton, G.J., Hajdu, J., Andersson, I., Schofield, C.J., and Baldwin, J.E. (1995). Crystal Structure of Isopenicillin N. Synthase. *Nature*, 375, 700–704.

The biosynthesis of most classes of antibacterial substances is regularly reviewed in *Natural Product Reports*. The most recent reports are by: O'Hagan, D. (1995) 12, 1; Herbert, R.B. (1995) 12, 55; and Dewick, P.M. (1995), 12, 101.

Abraham, E.P. (1985) In *Regulation of secondary-metabolite formation* (Kleinkauf, H., von Dohren, H., Dornauer, H. & Nesemann, G., eds) pp. 115–132, VCH Verlagsgesellschaft, Weinheim

Jensen, S.E. (1986) *C.R.C. Crit. Rev. Biotechnol.* 3, 277–301.

Nuesch, J., Heim, J. & Treichler, H.-J. (1987). *Ann. Rev. Microbiol.* 41, 51–75.

Chapter 5

Sutherland, R. (1991) Beta-lactamase inhibitors and reversal of antibiotic resistance. *Trends in Pharmacological Sciences*, 12, 227.

Finch, R.G., Hill, P. and Williams, P. (1995) Staphylococci: the emerging threat. *Chemistry and Industry*, 225.

Cohen, M.L. (1992) Epidemiology of drug resistance: implications for a post-antimicrobial era. *Science*, **257**, 1050.

Neu, H.C. (1992) The crisis in antibiotic resistance. *Science*, **257**, 1065.

Bottinger, E.C. (1994) Resistance to drugs targeting protein synthesis in bacteria. *Trends in Microbiology*, **2**, 416.

Cohen, M.L. (1994) Antimicrobial resistance. *Trends in Microbiology*, **2**, 422.

Sahm, D.F. and O'Brien, T.F. (1994) Detection and surveillance of antimicrobial resistance. *Trends in Microbiology*, **2**, 366.

Chapter 6

Roberts, S.M. (1984) Beta-lactams: past and present. *Chemistry and Industry*, 162.

Durckheimer, W., Blumbach, J., Lattrell, R. and Scheunemann, K.H. (1985) Recent developments in the field of beta-lactam antibiotics. *Angewandte Chemie*, **24**, 180.

Newall, C. (1987) Ceftazidime: an injectable antibiotic. *Chemistry in Britain*, 976.

Cimarusti, C.M. and Sykes, R.B. (1983) Monobactams: novel antibiotics. *Chemistry in Britain*, 302.

Corraz, A.J. *et al.* (1992) Dual-action penems and carbapenems. *J. Med. Chem.*, **35**, 1829.

Southgate, R. (1994) The synthesis of natural beta-lactam antibiotics. *Contemporary Organic Synthesis*, **1**, 417.

Grohe, K. (1992) Quinolone antibiotics: the new generation. *Chemistry in Britain*, 34.

Katz, L. and Donadio, S. (1993) Polyketide synthesis: prospects for hybrid antibiotics. *Annual Review of Microbiology*, **47**, 875.

Mann, J. (1995). Rules for the manipulation of polyketides. *Nature*, **375**, 533.

Marsh, J. and Goode, J.A. (1994) Antimicrobial peptides. *Ciba Foundation Symposium*, 186.

Rohr, J. Combinatorial biosynthesis—an approach in the near future (1995). *Angewandte Chemie*, **34**, 881–885.

Kao, C.M., Guanglin, L., Katz, L., Cane, D.E. and Khosla, C. (1994). Engineered biosynthesis of a triketide lactone. *J. Amer. Chem. Soc.*, **116**, 11612–11613.

Cortes, J., Wiesmann, K., Roberts, G.A., Brown, M.J., Staunton, J. and Leadlay, P.F. Repositioning the domain in a modular polyketide synthase. (1995). *Science*, **268**, 1487–1489.

Miller, M.J., Biswas, A. and Krook, M.A. (1983) A practical synthesis of monobactams. *Tetrahedron*, **39**, 2571.

Georg, G.I., Kant, J. and Gill, H.S. (1987) A synthetic approach to (+)-thienamycin. *J.Am.Chem.Soc.*, **109**, 1129.

Mathieu, L.G. and Sonea, S. (1995) A powerful bacterial world. *Trends Pharm.Sci.*, **16**, 112.

Berks, A.H. (1996) Synthesis of carbapenem antibiotics. *Tetrahedron*, **52**, 331–375.

Reviving the antibiotic miracle [several authors and articles]. *Science* (1994) **264**, 360.

Garrett, L. (1994) *The Coming Plague*. London, Virago.

Index